Sustainable Water Resources in the Built Environment

Sustainable Water Resources in the Built Environment

Marilyn Waite

Publishing
London • New York

Published by **IWA Publishing**
Alliance House
12 Caxton Street
London SW1H 0QS, UK
Telephone: +44 (0)20 7654 5500
Fax: +44 (0)20 7654 5555
Email: publications@iwap.co.uk
Web: www.iwapublishing.com

First published 2010

Typeset in India by OKS Prepress Services.

Cover design by www.designforpublishing.co.uk

British Library Cataloguing in Publication Data
A CIP catalogue record for this book is available from the British Library

Library of Congress Cataloging-in-Publication Data
A catalog record for this book is available from the Library of Congress

ISBN 10: 1843393239
ISBN 13: 9781843393238

Contents

ISBN: 9781843393238. Published by IWA Publishing, London, UK.

Acknowledgments

This book is the result of an extensive senior thesis conducted at Princeton University. I would like to foremost thank Professor Dr. Catherine Peters for all of her help and support in this project. I am greatly appreciative to the experts who provided me with specialized knowledge on the case studies in Kenya and Panama: Kiss and Cathcart Architects, members from the Mpala Research Centre community in Kenya, as well as members from the Smithsonian Tropical Research Institute in Panama.

The success of this entire work was dependent on field research in Kenya and Panama. I cannot express my gratitude for the funding provided by The Woodrow Wilson School of Public and International Affairs and The School of Engineering and Applied Sciences at Princeton University.

Thanks to all of those who have supported me throughout the writing process. There are numerous people who made this book possible—those who I called to ask a quick question, those who went the extra mile in assuring permissions, those who ensured that my stay in Kenya and Panama went smoothly, and others who supported me in many other capacities. I am grateful for your help.

ISBN: 9781843393238. Published by IWA Publishing, London, UK.

Acronyms and Abbreviations

ANAM	Autoridad Nacional del Ambiente [National Authority of the Environment]
APHA	American Public Health Association
BOD	Biochemical Oxygen Demand
CFU	Colony Forming Units
CU	Color Units
CW	Constructed Wetland
EAWAG	The Swiss Federal Institute for Environmental Science and Technology
EcoSan	Ecological Sanitation
EDL	Estimated Detection Limit
EC	European Community
E.coli	Escherichia coli
GI	Galvanized Iron
IDAAN	Instituto de Acueductos y Alcantarillados Nacionales [Institute of Aqueducts and National Sewage Systems]

ISBN: 9781843393238. Published by IWA Publishing, London, UK.

IDB	Inter-American Development Bank
IDRC	International Development Research Centre
IPCC	Intergovernmental Panel on Climate Change
Kshs	Kenyan Shillings [Currency]
MRC	Mpala Research Centre
NGO	Non-Governmental Organization
R	South African Rand [Currency]
Rs	Indian Rupee [Currency]
RWH	Rainwater Harvesting
STRI	Smithsonian Tropical Research Institute
SS	Suspended Solids
TSS	Total Suspended Solids
UNEP	United Nations Environment Programme
UN-Habitat	United Nations Human Settlements Program
USD	United States Dollar [Currency]
USDOE	United States Department of Energy
USEPA	United States Environmental Protection Agency
USGBC	United States Green Building Council
UTP	Universidad Tecnológica de Panamá [Technological University of Panama]
WHO	World Health Organization
WSP	Wastewater Stabilization Pond

Units

cu = cubic
ft = feet
g = gram
in = inches
km = kilometers
kVA = kilo-volt amperes
L = liters
m = meters
mg/L = milligrams per liter
mL = milliliter
mm = millimeters
sq = square
V = volt
°C = degrees Celsius
°F = degrees Fahrenheit

ISBN: 9781843393238. Published by IWA Publishing, London, UK.

Executive summary

This book addresses the critical water problems faced in developing countries. As much as 95 percent of the developing world's raw sewage is dumped directly into rivers, lakes, and oceans (Andersson, 2005). The polluted water, coupled with poor sanitation, cause the spread of diarrheal illnesses, hepatitis, intestinal worms, and other diseases. In Panama, 91% of the total population use improved drinking water sources and 72% of the total population use adequate sanitation facilities. In Kenya, 62% of the total population use improved drinking water sources and only 48% of the total population use adequate sanitation facilities. Because of the lack of adequate water and sanitation facilities in developing countries, it becomes imperative to search for ways in which to improve access to those basic necessities. Sustainable construction, through the use of water collecting and saving mechanisms such as rainwater harvesting, constructed wetlands, composting toilets, and low-flow fixtures, can provide at least a partial remedy to the water and sanitation problem in economically developing countries.

The focus is on low-cost sustainable water resources in the built environment. The water aspects of sustainable construction are analyzed in a technical and policy-oriented manner through two case studies: the Smithsonian Tropical

ISBN: 9781843393238. Published by IWA Publishing, London, UK.

Research Institute (STRI) building in Bocas del Toro, Panama and various buildings near the Mpala Research Centre (MRC) in Laikipia, Kenya.

This study supports the observation that rainwater harvested from a roof catchment surface can be in need of treatment before it is used for potable purposes; the rainwater from both the Kenya and Panama experiments were contaminated with coliforms. There were no major differences among the rainwater at different points in the STRI rainwater treatment system. Rainwater harvesting from six buildings in Laikipia, Kenya can meet domestic water needs for part of the year. A system of water rationing, perhaps by limiting rainwater collection to drinking consumption, will extend the rainwater storage volume for a longer period of time. From the interviews conducted in Kenya, there seems to be no major objections to the adoption of composting toilets; some prefer the Fossa-Alterna type of toilet while others prefer the Arbour-Loo type. Similarities and differences can be found in the case studies with respect to wastewater, potable water, budget, climate, culture, sustainable building usage, and water regulations.

A discussion of policies that encourage the use of low-cost sustainable water resources in the built environment is provided in the last chapter. Some of these policies include providing incentives for the use of composting toilets, having a combination of rainwater harvesting mandates, allowing tax incentives for new sustainable commercial and residential buildings, reducing the potential contamination in rainwater harvesting with a first-flush device, and including sustainable buildings as an integral part of national environmental policies.

The findings of this research suggest that sustainable building practices, such as on-site wastewater treatment through constructed wetlands and stabilization ponds, rainwater harvesting from roofs, and the use of composting toilets, can help to alleviate the water resources problems of inadequate water quantity, poor water quality, and site contamination by human waste.

1
Introduction
Sustainable Construction

1.1 INTRODUCTION: SUSTAINABLE CONSTRUCTION IN THE DEVELOPING WORLD

1.1.1 Sustainable construction

This book focuses on the water aspects of sustainable buildings in developing countries. Many environmental impacts of human activities are direct, immediate, and visible, while others take a long time to develop and may not be seen immediately after the action. Buildings are responsible for a large share of direct and indirect negative environmental impacts. In the United States alone, buildings account for 36% of total energy use, 65% of electricity consumption, 30% of waste output, and 12% of potable water consumption (USGBC, 2005). Developing countries can construct in a more sustainable manner as industrialization and

ISBN: 9781843393238. Published by IWA Publishing, London, UK.

urbanization increases. It is much easier to adopt facilities and methods that are low in pollution when in the process of development.

In order to prevent and curb the harmful impacts of construction, the concept of sustainability can be employed. Sustainable construction can be defined as the creation and responsible management of a healthy built environment based on resource efficiency and ecological principles. Sustainable buildings, or "green buildings," are designed, constructed, operated, and demolished in an environmentally and energy efficient manner (Mead, 2001). Green building concepts include water and energy efficiency and conservation, indoor environmental quality, resource and material efficiency, improved long term durability, and sustainable sites.

1.1.2 Benefits of sustainable construction

Sustainable buildings help to reduce the environmental impacts that construction and renovation can engender and can also have a positive bearing on economic and social conditions. The environmental benefits of sustainable construction include enhancing and protecting ecosystems and biodiversity, improving air and water quality, reducing solid waste, and conserving natural resources. The economic benefits include reducing operating costs, enhancing asset value and profits, improving employee productivity, and optimizing life-cycle economic performance (USGBC, 2005). In addition, many sustainability efforts have positive consequences for the eco-tourism industry, food security, health and sanitation, empowerment of women and children, and the general well-being of the public.

1.1.3 Water aspects of sustainable construction

One critical aspect of sustainable construction is water collection, use, and reuse. Two key factors determine appropriate water use: good *quality* and adequate *quantity*. As stated by Avijit Gupta in *Ecology and Development in the Third World*, "any development that alters the local environment [such as the construction or renovation of a building] alters the paths, and thereby the available amount and quality of the water supply" (Gupta, 1998). Currently, most rural communities in developing countries have a water demand that far exceeds the water supply, therefore creating a situation of endless water deficiency (Onyenechere, 2004). In addition, many local waterways are heavily polluted from numerous sources, including direct discharge of excreta, debris and pollution from construction, as well as runoff from industrial and agricultural activities.

The water available for human use (from the atmosphere, land surface, and shallow subsurface) constitutes an extremely small part of the total world inventory. This world inventory includes water from the oceans, ice sheets, and deep underground. Although it is technologically possible to extract this water for human needs such as drinking, bathing, and cooking, it is expensive and uneconomical, especially for developing nations. World water supply is generally based on precipitation (and secondarily groundwater). However, there is an uneven distribution of precipitation among continents, which adds to the problem of disproportionate water stress (ratio between human water withdrawals and water availability) (Lozán, Grassl, Hupfer, Menzel and Schönwiese, 2004). In arid climate regions, potential evaporation is larger than precipitation (Falkenmark, 2007).

Figure 1.1 and Tables 1.1 to 1.2 show the hydrological cycle, the breakdown of the total world inventory of water, and a list of types of water use. The amount of water that goes through the different paths of the hydrological cycle is determined by local conditions and varies seasonally. Thus, when designing a green building to satisfy the water needs of a particular area, the local climate and geography are important. For example, area rainfall will determine how much rain can be harvested from a roof, how water might be reused to support agriculture, and what kinds of water conservation tools should be built into the structure. Our water supply is completely dependent on the hydrological cycle. One must bear this in mind when accessing water availability for various types of water use. The global average water footprint (total volume of freshwater used to produce the goods and services consumed by people), is about 1,240 m^3 per capita per year, with wide variations among nations (Hoekstra and Chapagain, 2007).

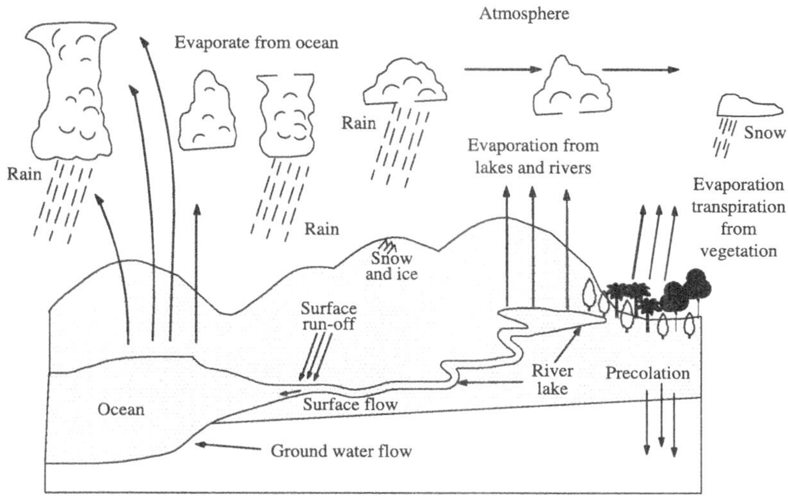

Figure 1.1 The hydrological cycle

Table 1.1 Water storage (Gupta, 1998) from: Ecology and development in the third world, A. Gupta, Copyright 1998 Taylor & Francis. Reproduced by permission of Taylor & Francis Books, UK

Oceans: 97%	Atmosphere: 0.035%	Ice Sheets and Glaciers: 75% of all **Fresh** Water
	Rivers: 0.03% Lakes: 0.3% Moisture in Soil: 0.06% Groundwater (<750 m depth): 11% Groundwater (>750 m depth): 14% *Percentages in this column refer to **Fresh** Water Distribution	

Table 1.2 Types of water use (Gupta, 1998) from: Ecology and development in the third world, A. Gupta, Copyright 1998 Taylor & Francis. Reproduced by permission of Taylor & Francis Books, UK

Type of use	Characteristics
Municipal: potable water	Consumptive.
Municipal: other uses	Usually heavily polluted after use. Possibility of reuse after proper treatment.
Industrial	Various uses, a large proportion for cooling purposes. Level of pollution depends on industry involved. Substantial portion could be reused after treatment.
Hydro-electricity generation	Almost entirely reusable. Some evaporation loss from reservoirs.
Agriculture	Polluted by fertilizers, pesticides, and saline concentration. Could be reused under favourable conditions after treatment. Otherwise usually ends up in local streams.
Navigation	All reusable. Level of pollution low.
Recreation	Reusable. Level of pollution low. Some pathogenic organisms might be present.
Propagation of fish and wildlife	Reusable. Level of pollution low.

1.1.4 Critical situation in developing countries

As the population of developing countries continues to increase, so will human activities that have the potential of harming the environment. In order to provide food, clothing, *shelter*, education, and employment for a growing population, human activities will have to increase significantly (Biswas, Khoshoo and Khosla, 1990). In addition to meeting the needs of a *growing* population, some of the current populations of developing nations are undergoing massive development.

The environmental degradation engendered by development can be troublesome. As Asit K. Biswas, professor and international consultant on water resources, stated, "the real challenge facing decision-makers in developing countries is one of providing needs of a continually expanding population without destroying the environment and the resource base on which the very process of development depends" (Biswas, Khoshoo and Khosla, 1990). Environmental degradation in developing countries can be attributed to pollution and toxic waste disposal problems from industrialization, increased pressure on the environment due to activities such as farming of marginal lands and gathering of fuelwood by the poor, and the extraction of natural resources for either national consumption or use in economically developed countries (Gupta, 1998). However, in addition to those problems and others, "development" can create environmental degradation through the negative consequences of building construction.

1.2 OBJECTIVE AND SCOPE

1.2.1 Objectives

This book specifically examines two case studies in Panama and Kenya. The Panama study was completed at the Smithsonian Tropical Research Institute (STRI) building in the province of Bocas del Toro, on the island of Isla Colón, and in the town of Bocas del Toro, Panama. The Kenya case study was completed in the area surrounding the village of the Mpala Research Centre (MRC), which is located in the Central Province, in the Laikipia District of Kenya. The objectives are (1) to provide technical analyses of the water aspects of sustainable buildings in the two case studies in order to provide technical support for the feasibility of green buildings and (2) to consider the policy implications of these sustainable practices. The specific objectives are seven-fold:

(1) to determine if rainwater harvested from a roof catchment surface is in need of treatment before it is used for potable purposes
(2) to determine the quality of the rainwater at different points in the STRI rainwater harvesting (RWH) system
(3) to illustrate the potential for sustainable water practices such as constructed wetlands and RWH in addressing water management problems in areas like Bocas del Toro, Panama
(4) to determine if RWH from village homes and schools in Kenya can meet the water demand and whether or not the harvested water would need treatment

(5) to find out if composting toilets are feasible near MRC given the constraints of cultural acceptability and economic use of the compost
(6) to analyze the similarities and differences between the two case studies in order to draw general conclusions about the main issues that should be considered when incorporating water saving mechanisms in buildings in wet and dry regions
(7) to discuss policies that encourage the use of sustainable buildings

The overall goal is to promote the use of sustainable buildings as an avenue to help protect and enhance the environment (physically and socially), improve access to safe drinking water and sanitation, and enhance the local economy. It is hoped that this work will create awareness among readers on (1) the potential for water supply and savings within sustainable buildings in developing countries and on (2) a sense of the different environmental, social, political, geographical, and other factors that influence water resources in the built environment in developing countries.

1.2.2 Overview of approach

The approach taken to meet the aforementioned objectives combines the investigation of relevant literature and field work. A rainwater harvesting system and a wastewater stabilization pond/constructed wetland system were examined at the STRI building in Bocas del Toro, Panama. At STRI, water quality tests on the rainwater system was performed to determine suitability for drinking and appropriate treatment options. In addition, information regarding the treatment of wastewater was analyzed using a natural plant system, the water conservation techniques incorporated into the building, and the details of the rainwater harvesting system. The possibility for the incorporation of rainwater harvesting and composting toilets to various buildings near MRC in Kenya was examined. Water quality tests were also conducted on the rainwater systems in place, as well as the local river water, to determine suitability for drinking and appropriate treatment options. Rainwater harvesting system components were designed for a range of buildings that was analyzed in Kenya.

1.2.3 Case Studies in Kenya and Panama

The Kenya and Panama case studies represent (1) different parts of the developing world with different physical environments and cultural norms and (2) different budgetary constraints that can arise in regard to sustainable buildings. Panama is a country with generally abundant rainfall, while Kenya is often plagued with droughts. The Panama case study represents a situation in which a developed nation or international research group builds a research center in a developing nation. Buildings in this scenario tend to have budgets appropriate to medium to high-cost construction methods and technologies. For example, STRI in Bocas can afford to use photovoltaic technology to provide solar energy. The case study in Kenya, on the other hand, represents a situation in which the houses and schools of local inhabitants in developing countries can be built in a sustainable and affordable manner. Buildings in this scenario tend to have budgets appropriate to low-cost construction methods and technologies. In both situations, the sustainable building can incorporate water conservation, safe water quality, and the overall protection and enhancement of the environment, while also positively impacting the local community socially and economically.

1.3 CURRENT STATE OF WATER RESOURCES

The state of water resources and sanitation is precarious. Because of the lack of adequate water and sanitation in developing countries, it becomes imperative to search for ways in which to improve access to those basic necessities. As much as 95 percent of the developing world's raw sewage is dumped directly into rivers, lakes, and oceans (Andersson, 2005). The polluted water coupled with poor sanitation causes the spread of diarrheal illnesses, hepatitis, intestinal worms, and other diseases. One of the main sources of protein and income—fisheries—are damaged by untreated waste. In addition, there is not enough money or water to meet the global demand for the conventional Western sanitation method of using water to transport waste away from toilets. By 2025, it is expected that 35% of the world's population will face *severe* water shortages (Andersson, 2005). Water-intensive sanitation systems will be unfeasible in the near future. There is a widely accepted water need value of 50 L/person/day for domestic water supply, often considered as a human right; in addition, the water required to produce a nutritionally acceptable diet for one person at 3,000 kcal/person/day requires 70 times as much water on a per capita basis that the 50 L standard (Falkenmark, 2007).

1.4 WATER RESOURCES AND CLIMATE CHANGE

With global population expected to rise to 8.7 billion by 2050 (Bates, Kundzewicz, Wu and Palutikof, 2008) there will be an increasing need for reliable and safe water for potable and non-potable purposes. At the same time, the climate change outlook for 2050 is not encouraging: more extreme weather events, decreasing land precipitation in some of the already drought-prone regions, increasing temperatures, increasing water stress, etc. According to the Intergovernmental Panel on Climate Change (IPCC), point-source discharges of nutrients, heavy metals and organic substances are likely to increase in developing countries; higher water temperatures, increased precipitation intensity, and longer periods of low flow are projected to worsen many forms of water pollution (Bates, Kundzewicz, Wu and Palutikof, 2008).

The impacts of climate change on the water cycle are both direct and indirect. Direct effects could mean dryer regions, serious increases in floods, regions whose rainfall exceed the carrying capacity of the water channels, variations in water storage and fluxes (on surface land and surface water of lakes, in soil moisture, groundwater, glaciers, snow packs, wetlands, reservoirs, etc.) (UNESCO-WWAP, 2009). Indirect effects include climate change's potential to rapidly increase human migration. This in turn, impacts water systems designed for less demand.

The link between energy and water, also known as the energy-water nexus, is an important factor when considering sustainable water resources in the built environment. Water is often needed for energy creation (raw material processing, extraction, cooling, etc.) and energy is often needed for the collection, distribution and purification of water. In addition to being intricately linked, energy and water resources are also at the cornerstone of climate change. Our current use of energy leads to the emission of greenhouse gases that contribute to climate change. This change in climate, in turn, leads to drastic changes in water resources. The impacts of anthropogenic climate change on water resources have been well-documented by the IPCC. They include changes to:

- the hydrological cycle
  - increase in atmospheric water vapor content
  - change in precipitation patterns
  - increase in frequency of heavy precipitation events
  - reduction in snow cover
  - widespread melting of ice
  - change in soil moisture and runoff
- dry climates and wet climates
  - decrease in water resources due to climate change in many semi-arid and arid areas including southern Africa
  - increase of annual average river runoff and water availability at high latitudes and in some wet tropical areas

- water stress
  - doubling of the area of land subject to increasing water stress due to climate change by 2050 (Bates, Kundzewicz, Wu and Palutikof, 2008).

The two environments explored in this book, a dry-climate region of Kenya and a wet-climate region of Panama, are thus particularly interesting because of the various and differing climate change implications. According to the IPCC, (1) droughts have become more common, especially in the tropics and sub-tropics, since the 1970s, (2) it is likely that the area affected by drought has increased since the 1970s, and (3) it is more likely than not that there is a human contribution to this trend (Bates, Kundzewicz, Wu and Palutikof, 2008). The case study in Kenya takes place in a drought-prone area that may fall into this category.

Both countries considered in this book have less than perfect drinking water sources and sanitation facilities. In Panama, while 99% of the urban population has access to improved drinking water sources, only 79% of the rural population has access to improved drinking water sources (UNICEF, 2002b). *In Panama, 91% of the total population uses improved drinking water sources.* The United Nations Children's Fund (UNICEF) estimates that 89% of the urban population uses adequate sanitation facilities, while only 51% of the rural population uses adequate sanitation facilities (UNICEF, 2002b). *In Panama, 72% of the total population uses adequate sanitation facilities.* The water and sanitation problem is more acute in Kenya, where 89% of the urban population is using improved drinking water sources, but only 46% of the rural population is using improved drinking water sources. *In Kenya, 62% of the total population uses improved drinking water sources.* In regards to adequate sanitation, only 56% of the urban population and 43% of the rural population is using adequate sanitation facilities. *In Kenya, only 48% of the total population uses adequate sanitation facilities.*

The World Health Organization (WHO) and UNICEF Joint Monitoring Programme (JMP) for Water Supply and Sanitation defines adequate sanitation in terms of technology used. Facilities that are not shared and that hygienically separate human excreta from human contact are considered adequate (WHO and UNICEF JMP, 2005). Certain technologies are more likely than others to meet these adequacy standards. The JMP calls technologies that meet adequacy standards "improved" and those that do not, "not improved." It is assumed by the JMP that people using certain types of drinking water sources, delivery or extraction technologies, have access to drinking water of an acceptable quality to meet their basic health needs. Table 1.3 shows how improved drinking water supply and sanitation facilities are defined.

Table 1.3 Classifications of improved and not-improved drinking water supply and sanitation facilities (WHO and UNICEF JMP, 2005)

	Improved	Not Improved
Drinking Water Supply	Piped water into dwelling, plot or yard Public tap/standpipe Tubewell/Borehole Protected dug well Protected spring Rainwater collection	Unprotected dug well Unprotected spring Cart with small tank/drum Bottled water* Tanker-truck Surface water (river, dam, lake, pond, stream, canal, irrigation channels)
Sanitation**	Flush/pour flush to: – piped sewer system – septic tank – pit latrine – unknown place/not know where Ventilated Improved Pit latrine (VIP) Pit latrine with slab Composting toilet	Flush/Pour flush to: – elsewhere Pit latrine without slab/open pit Bucket Hanging toilet/hanging latrine No facilities or bush or field

*Bottled water is considered improved when users have access to another improved source as well
**Only private facilities are considered improved

Sustainable construction, through the use of water collecting and saving mechanisms such as rainwater harvesting, constructed wetlands, composting toilets, and low-flow fixtures, can provide at least a partial remedy to the water and sanitation problem in developing countries such as Panama and Kenya.

1.5 ROADMAP

The pages that follow offer detailed information regarding the two case studies. There is a background discussion on rainwater harvesting in developing countries, water quality indicators, wastewater treatment and reuse schemes such as constructed wetlands, and water conservation tools such as composting eco-toilets. For the *Panama* case study, the following is given: a detailed description of the STRI site and water management practices of the surrounding community, a detailed description of the water aspects of the STRI building, and the water quality experimental methods, results, and analysis. For the Kenya case study, the following is provided: a detailed description of the MRC and surrounding community sites and water management practices, the water quality experimental methods, results, and analysis, and designs for RWH systems for various buildings near MRC. Finally, an analysis of the policy implications of the water aspects of sustainable construction in Panama and Kenya based on relevant country water policies is given.

2
Background
Water aspects of sustainable buildings

2.1 INTRODUCTION: WATER ASPECTS OF SUSTAINABLE BUILDINGS

This chapter presents background information on the potable water aspects of sustainable buildings in developing countries, namely domestic rainwater harvesting (RWH). Specifically, this chapter will cover a description of water quality indicators used in the two case study experiments, rainwater harvesting sources of contamination and methods of prevention and treatment, advantages and disadvantages to rainwater harvesting, and structural and economic aspects of RWH.

ISBN: 9781843393238. Published by IWA Publishing, London, UK.

2.2 WATER QUALITY INDICATORS

2.2.1 Description of indicators

This section will describe some of the water quality parameters used to test water for potable purposes. The following parameters are discussed: total suspended solids (TSS), turbidity, total and fecal coliform, sulfates, pH, color, nitrates, and nitrites.

Total solids are matter suspended or dissolved in water; it includes both total suspended solids (TSS) and total dissolved solids (TDS). TSS are solids that can be trapped by a filter. TSS is primarily a concern for surface water; it decreases water clarity, which affects the ability of fish to see and catch food and reduces rates of photosynthesis (which causes a decrease in dissolved oxygen levels). In drinking water, TSS does not pose a major threat; however, it can indicate that there are various concentrations of bacteria, nutrients, pesticides, and metals in the water. Turbidity is closely related to TSS. Turbidity is a measure of the amount of light scattered and absorbed by water because of the suspended matter in the water. Turbidity may be due to a single foreign substance in water, or due to a mixture of many substances. These particles may range in size from fine colloidal materials to coarse grains of sand. Turbidity is not known to have any health effects, but it can indicate the presence of inorganic substances such as clay, rock flour, silt, calcium carbonate, silica, iron, manganese, sulfur, or industrial wastes, organic substances such as various microorganisms, finely divided animal matter, and bacteria and other disease-causing organisms.

Coliform bacteria are members of the family Enterobacteriaceae. Most coliforms are not pathogenic, but they serve as indicators of the microbial quality of the water. Total coliforms are mostly natural residents of soil and water. Escherichia coli (E. coli) is a genus (Escherichia) and species (coli) in the total coliform group. It is also a fecal coliform (FC), a smaller subset of total coliforms. E. coli is a commonly used indicator of fecal contamination. It is present in high numbers in the gastrointestinal tracts of warm-blooded animals (such as dogs, cats, birds, mammals, and people) and thus provides a sensitive measure of fecal contamination. Fecal coliforms are defined as coliforms that can grow at 44.5°C, which is too hot for most of the other coliforms. Water contaminated with fecal coliform may cause diarrhea, nausea, headaches, and other symptoms (USEPA, 2005a). In some people, especially the elderly and children of age five years and below, E. coli infection (from strain 0157) can cause hemolytic uremic syndrome, in which the red blood cells are destroyed and the kidneys fail.

Color, sulfate, and pH, are listed as secondary drinking water regulations by the United States Environmental Protection Agency (USEPA). They are not considered to present a risk to human health at the maximum contamination level set by the USEPA. The problems with secondary drinking water contaminants normally arise from their aesthetic effects (unpleasant tastes and odors). The color of the water is affected by, among other aspects, small microscopic particles, inorganic contaminants such as metals, colloidal suspensions, noncolloidal organic acids, and neutral salts. Color may be indicative of dissolved organic material, inadequate treatment, high disinfectant demand, and the potential for the production of excess amounts of disinfectant by-products. Sulfate, a naturally-occurring substance in drinking water, creates a salty taste if it has a high concentration. Health concerns regarding sulfate in drinking water have been raised because of reports that diarrhea may be associated with the ingestion of water containing high levels of sulfate (USEPA, 2005a). The pH of water reflects how acidic or basic it is. It stands for "potential hydrogen," which is a measure of the amount of free hydrogen in the water. A pH of 7 is "neutral"; it indicates that there is no acid or alkalinity present; this is the pH of pure water. A pH below 7 indicates the presence of acid and a pH above 7 indicates alkalinity. Water with a low pH can be acidic, soft, and corrosive; water with a high pH can indicate that it is hard, which can cause aesthetic problems such as an alkali taste.

Nitrates and nitrites are nitrogen-oxygen chemical units that combine with various organic and inorganic compounds (USEPA, 2005a). Nitrates are converted into nitrites after consumption. Excessive nitrate/nitrite levels in drinking water can have fatal effects. From a lifetime of exposure to high levels of nitrates and nitrites, diuresis, increased starchy deposits, and haemorrhaging of the spleen can result. Methemoglobinemia is a major concern of high nitrate and nitrite levels in water. Methemoglobinemia is a blood disorder caused when nitrite interacts with the hemoglobin in red blood cells. Unlike hemoglobin, the methemoglobin formed in this interaction cannot carry sufficient oxygen to the body's cells and tissues. This disorder, also known as "blue baby syndrome," impacts mostly infants, and can cause shortness of breath, blueness of the skin, and death.

2.2.2 Water quality standards

International organizations, country governments, and various governing groups have developed water quality standards for drinking water. These standards are set so that water-borne illnesses and related diseases are eliminated or reduced. The World Health Organization (WHO) produces international norms on water quality and in the form of guidelines that are used as the basis for regulation and standard setting in countries all over the world. Many governments and inter-governmental entities, such as the United States Environmental Protection Agency (USEPA) and the European Community (EC), have set their own water quality regulations along with penalties of not meeting those requirements. Table 2.1 presents the WHO, EPA, and EC standards for the parameters tested in the water quality experiments (See Chapter 5).

Table 2.1 Drinking water quality standards

Parameter	Unit*	WHO Guideline	USEPA Max Contamination Level**	EC Max Admissible Concentration***
pH	pH units	6.5–8.5	6.5–8.5 ◊	6.5–8.5
TSS	mg/L	NS	NS	NS
Turbidity	NTU	5	1 ○	4 FTU
Fecal Coliform	**CFU/100 mL**	**0**	**<1** □	**0**
Total Coliform	**CFU/100 mL**	**0**	**<1** □	**0**
Sulfates	mg/L	250	250 ◊	250
Nitrates	mg/L	50	10	50
Nitrites	mg/L	3	1	0.1
Color	Cu	15	NS	20 mg Pt-Co/L

NS No Standard
* *Unit Unless Otherwise Noted*
** *United States Environmental Protection Agency*
*** *European Community (established by European Committee for Environmental Legislation)*

◊ United States National Secondary Drinking Water Regulations
○ The turbidity limit is 1 NTU at any given time; the turbidity is not to exceed 0.3 NTU in 95% of water samples for a given month
□ No more than 5% of water samples can test positive for the presence of total coliform in a given month
If total coliforms are present, fecal coliforms are tested to determine presence/absence
Absence is declared if there are less than 1 CFU/100 mL; presence is declared if there are equal to or greater than 1 CFU/100 mL

WHO Guideline	Found in WHO's Third Edition of Guidelines for Drinking Water Quality (WHO, 2005), (WHO, 2006), (Hach, 2003), (Rooney, 2005)
USEPA Standards	Found in USEPA's List of Drinking Water Contaminants (USEPA, 2005), (USEPA, 2009)
EC Standards	Found in the EC's Drinking Water Directive (European Commission, 2004)

2.3 RAINWATER HARVESTING IN DEVELOPING COUNTRIES

Taking advantage of rainfall to serve one's water needs is an important element of sustainable buildings. The widespread practices of exploiting surface water and groundwater sources are often inadequate and damaging to the environment. For example, pumping groundwater disrupts the water table and requires energy and high costs for abstraction as the water table becomes lower. Furthermore, groundwater can contain harmful chemical pollutants such as arsenic. Other dangers of overexploitation of groundwater resources include plastic deformation of the aquifers, intrusion of sea water in areas near the sea, and the reduction of storage capacity of underground aquifers which leads to cessation of streams (IEA and SID, 2001). Surface waters can be heavily polluted, lack adequate volume to satisfy the water demand, and can be otherwise inappropriate for water collection. In Kenya and Panama, many surface waters are contaminated and inadequate for drinking and swimming. Therefore, rainwater harvesting (RWH), or the collection of rainwater for various uses, is a viable water source alternative. This is especially important to resource-limited areas, where surface water quality may be poor, where other extraction methods may be too difficult or too expensive, and where water may be scarce.

In the Kenya case study, RWH is a crucial component given the lack of rainfall and adequate water resources. Tables 2.2 and 2.3 show the critical water deficits in selected provinces in Kenya as well as the water quality from 1963–2000.

Table 2.2 Water deficits in Kenya (Mwagore, 2002)

Province	Water deficit (m^3/day)
North-Eastern	86,020
Eastern	281,516
Coast	76,241
Rift Valley	197,544
Central	63,200

The total water deficit in Kenya is 705,000 m^3/day. The following calculations will illustrate the rainwater harvesting needed to curb the deficit in Kenya assuming that corrugated galvanized iron (GI) roofs are employed and the mean annual rainfall is 567 mm (IEA and SID, 2001). The equations used to calculate how much water can be supplied daily from RWH are as follows:

$$\begin{aligned} \textit{Gross water supply} &= (\textit{area of roof horizontal projection}) \\ &\quad \times (\textit{annual rainfall}) \end{aligned} \tag{2.1}$$

$$\textit{Net water supply} = C_R \times (\textit{gross water supply}) \tag{2.2}$$

$$\frac{\textit{Net water supply}}{\textit{Number of days in a year}} = \textit{daily water storage volume} \tag{2.3}$$

$$1\ \mathrm{m}^3 = 1000\,\textit{liters}\,(L) \tag{2.4}$$

where,

Gross water supply (yearly) is in m^3
Surface area is in m^2
Rainfall (yearly) is in m
C_R is the runoff coefficient for roof material [0.9 for corrugated GI roofs]
Number of days in a year is 365.25

Table 2.3 Water quality in Kenya, 1963–2000 (Mwagore, 2002)

		Actual level				
Parameter	Desired level	1963	1974	1984	1999	2000
Colour	colourless	colourless	colourless	Slightly coloured	Coloured turbid	Highly coloured
Turbidity Sediment Load	Clear nil	Clear nil	Clear slight	Turbid moderate	heavy	Very turbid heavy
Total Dissolved Solids	Normal	normal	Fair	Moderate increase	Moderate increase	Very high increase
Bacteriological Contamination	No pathogenic bacteria	little contamin.	little contamin.	little contamin.	gross contamin.	gross contamin.
Biological oxygen demand	Less than 5 mg/l	5–20	10–20	20–40	40–60	60+

To find out the total area needed to meet the water deficit, equations 2.1 through 2.4 are applied as follows:

$$\text{Net water supply needed (water deficit)} = \frac{C_R \times (total\ surface\ area) \times (rainfall)}{365.25\, days}$$

$$\text{Total surface area} = \frac{704,521\, m^3/day \times (365.25\, days)}{0.9 \times 0.567\, m}$$

$$\text{Total surface area} = 504,000,000\, \text{m}^2$$

Thus the total area of RWH needed to meet water deficit in Kenya is about 504 million m^2. This is technically feasible since the land area in Kenya is about 569 billion m^2. The next question asked is how many roofs of size 30 m^2 each (average household roof size in Kenya case study) would be needed to meet the water deficit in Kenya. To answer this, equations 2.1 through 2.4 are again employed.

Net water supply needed (water deficit)

$$= \frac{C_R \times (surface\ area\ of\ roof) \times (rainfall) \times (\text{number of households/number of roofs})}{365.25\, days}$$

Number of households (number of roofs)

$$= \frac{(704,521\, \text{m}^3/\text{day})/(0.9 \times 30\, \text{m}^2 \times 0.567\, \text{m})}{365.25\, days}$$

Number of households (number of roofs) = 17,000,000 households

Thus the total number of households needed to meet the water deficit in Kenya is about 17 million, assuming the roofs are 30 m^2. There are approximately 30 million people in Kenya; assuming the average household contains 4 people, there are 7.5 million household size units in Kenya. Thus the requirement for households to collect rainwater in order to meet the water deficit is unlikely. This indicates that RWH can serve as a partial solution to the water shortages in Kenya, but not a full and complete remedy. It is worth noting that there are a great deal of scenarios in which RWH from buildings can help to curb the water deficit, including constructing larger buildings with larger roofs, collecting water from school roofs, and collecting water from other public and private buildings.

Rainwater harvesting can be incorporated into the design of a building through different roof catchment systems. The impact on water quality from various roofs depends on the type of roof material, climatic conditions, and the surrounding environment (Thomas & Greene, 1993). Figures 2.1 and 2.2 provide a basic schematic diagram of RWH from a roof, as well as an example of a poorly-maintained roof. Roofs made out of metal or tile are the easiest to configure, and they provide the cleanest water; corrugated galvanized iron roofs are prevalent in many humid and sub-humid regions, but tend to rust in coastal areas (Thomas & Greene, 1993). Materials commonly used for roof catchments are corrugated aluminum and galvanized iron, concrete, fiberglass shingles, tiles, slates, mud, bamboo, and asbestos. Some of these materials can create health hazards due to their toxicity when aging or degrading (such as asbestos), or their ability to easily become infected with termites, fungus, and bacteria (UNEP, 2002b). Diffuse sources of pollution from the atmosphere, the feces of birds and other animals, and other particulate matter can contaminate the water by falling onto the roof.

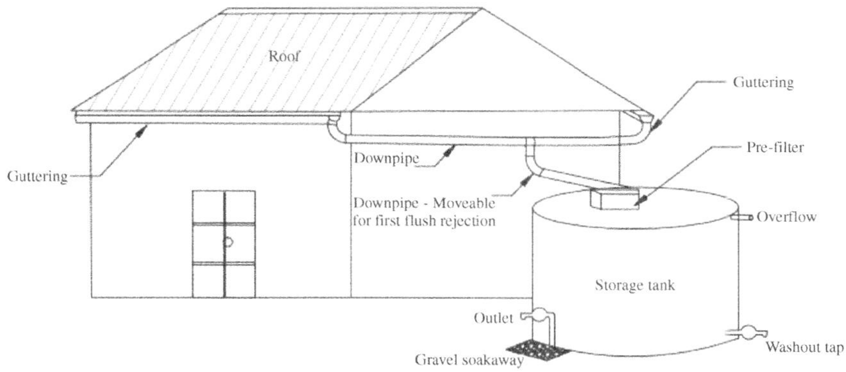

Figure 2.1 Typical very low cost roofwater harvesting system in a developing country (Turner, 2000)

2.3.1 Biological contamination

Biological contamination of rainwater can be grouped into two categories: insect vector breeding and microbiological contamination.

2.3.1.1 Insect vector breeding

There are currently no reported cases of illness due to insect breeding in rainwater storage tanks. However, there are several reports of mosquitoes and other insects breeding in rainwater tanks. The health implications of disease-carrying and disease-causing insects breeding in stored rainwater are cause for concern. However, there are methods of prevention and treatment of the rainwater to eliminate this threat. For example, there are methods that prevent physical access of mosquitoes (such as mosquito netting), as well as chemical and biological additives that prevent insect breeding.

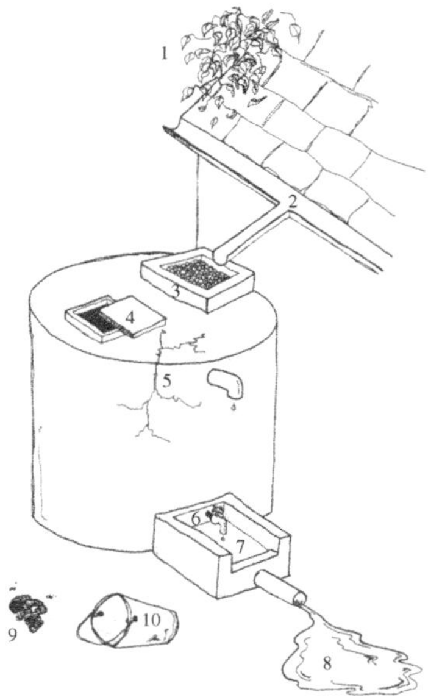

1 - Contamination on roof catchment area
2 - Dirty guttering channels
3 - Deficiency in filter box at tank inlet
4 - Point of entry not properly covered
5 - Cracks
6 - Defective tap
7 - Floor under tap defective or dirty
8 - Water collection area not drained
9 - Pollution around tank
10 - Bucket left in a place where it can easily become contaminated

Figure 2.2 A poorly maintained rainwater tank (WHO, 1997)

2.3.1.2 Microbiological contamination

Microbiological contamination of rainwater can occur through catchment surface or while it is in storage. The introduction of microbiological pollution can occur through intermediate vector such as insects, lizards, and other small animals that enter and die in the storage tank. High levels of bacterial contamination can occur through dust, organic matter, and the fecal material of birds and small mammals on the catchment surface. There have been several reports in many countries, which have stated that rainwater supplies have not met the required microbiological standards set by the World Health Organization (WHO) and other national drinking water quality standards.

2.3.1.3 Chemical contamination

Sources of chemical contamination can be divided into two types—those arising from off-site sources and those arising from on-site sources. Off-site sources include urban traffic, industrial emissions, and poor agricultural practices. Soil from industrial, mining or agricultural areas may contain arsenic, heavy metals, pesticide residues or other chemicals. If dust is blown onto a rainwater collection roof, those chemicals in the soil can be washed into the storage tank. One agricultural practice that can lead to chemical contamination of the rainwater entering a storage tank is the aerial application of fertilizers and pesticides, known as crop dusting. Urban areas are especially troublesome because of their climatic conditions (see Table 2.4) and the chemical compounds present in the air (see Table 2.5). On-site sources are those in the immediate vicinity of the storage tank. These sources include characteristics of the roof catchment, materials used in construction of the roof, gutters, piping, and tanks.

2.3.2 Reducing contamination

Research has revealed that bacterial and chemical contamination can be prevented or reduced through introduction of filters and a first flush system. The *"first flush" is defined as the first flow of rainwater after a dry period, which tends to contain contaminants from the roof.* Bacterial contamination can be minimized by (1) constantly cleaning rooftop surfaces and gutters, (2) using an installed first flush device which directs the initial batch of rainwater away from the tank to prevent the entry of debris from the catchment area into the storage container, (3) checking and cleaning the storage tank periodically with chlorine solution followed by rinsing, (4) maintaining gutters and downpipes to eliminate leaks, (5) using fully closed tanks to prevent evaporation, the growth of algae, and contamination, and (6) having all tank openings screened to prevent the access of mosquitoes, insects, and rodents (UNEP, 1999). Chemical contamination from off-site sources can be reduced by decreasing harmful chemical emissions from industrial, traffic, and agricultural sectors. Automobiles that use unleaded gasoline help prevent lead contamination. Chemical contamination from on-site sources can be minimized by also not collecting rainwater from roofs coated with bitumen-based materials or painted with lead-based paints (Australian Government, 2004).

In the case of the contamination of rainwater, there are several methods of treatment that are currently being employed. These include boiling, solar disinfection, chlorination, and filtration.

Table 2.4 Local climatic alterations produced by cities (Gupta, 1998) From: Ecology and Development in the Third World, A. Gupta, Copyright 1998 Taylor & Francis. Reproduced by permission of Taylor & Francis Books UK

Elements	*Compared to rural environs*
Contaminants:	
Condensation nuclei	10–100 times more
Particulats (dust)	10–50 times more
Gaseous admixtures	5–25 times more
Radiation:	
Total on horizontal surface	10–20% less
Ultraviolet, low sun	30% less
Ultraviolet, high sun	5% less
Sunshine duration	5–15% less
Cloudiness:	
Clouds	5–10% more
Fog, winter	100% more
Fog, summer	20–30% more
Precipitation:	
Amounts	5–10% more
Days with <5 mm	10% more
Temperature:	
Annual mean	0.5–1.0°C more
Winter minima (average)	1–2°C more
Surface relative humidity:	
Annual mean	6% less
Winter	2% less
Summer	8% less
Wind speed:	
Annual mean	20–30% less
Extreme gusts	10–20% less
Calm	5–20% more

Table 2.5 Concentrations of some air pollutants in city atmospheres (in parts per million by volume) (Gupta, 1998) From: Ecology and Development in the Third World, A. Gupta, Copyright 1998 Taylor & Francis. Reproduced by permission of Taylor & Francis Books UK

Carbon dioxide	300–1,000
Carbon monoxide	1–200
Sulphur dioxide	0.01–3
Oxides of nitrogen	0.01–1
Aldehydes	0.01–1
Oxidants, including ozone	0.00–0.8
Chlorides	0.00–0.3
Ammonia	0.00–0.21

Note: This happens with a 10–100 times increase in condensation nuclei, and 10–50 times increase in particulates.

2.3.3 Examples

In an Australian study by P.R. Thomas and G.R. Greene, samples of rainwater were collected from rural, urban, and industrial areas using galvanized iron roofs and concrete tile roofs. The collected samples were analyzed for pH, conductivity, turbidity, suspended solids, fecal coliform, nitrates, zinc, and lead using American Public Health Association (APHA) Standard Methods. All of the samples tested were not in the norm of the WHO drinking water guidelines (Thomas & Greene, 1993). However, the control samples collected from three open areas satisfied the WHO guidelines. Most of the samples *showed the presence of a bacterial load,* indicating that some purification method was needed in order to use the water for domestic consumption. They found that (1) the rainwater quality decreased with the increase in number of dry days between rainfall events, (2) urban and industrial areas were characterized by higher levels of lead from motor vehicle emissions, and (3) rural area roof catchments carried increased levels of nitrates due to agricultural activities.

In a study conducted in Engugu, Nigeria, it was proposed that large scale rainwater harvesting was needed to cease the problem of water shortage. Like many places, this area of Nigeria is faced with the problem of rainfall seasonality. Most of the rural dwellers can only afford small unit containers like pots to store water (Onyenechere, 2004). By the time the dry season advances, all of the water stored from the rain is finished. Also like many places in the developing world, houses with thatched roofs are prevalent in three of the Engugu communities surveyed. These thatched roofs have not been shown to be effective for rainwater harvesting. Thus, the inhabitants of the rural communities place rainwater collection pots below banana or coconut trees. It was recommended by E.C. Onyenechere, the author of the study, (1) that rainwater collection tanks be implemented, (2) that the tanks should prevent and reduce contamination through the use of mosquito screens and covers, and (3) that the tanks be fully closed.

2.3.4 Advantages and disadvantages of RWH

Table 2.6 shows the advantages and disadvantages of RWH in comparison to other water sources.

Table 2.6 Advantages and disadvantages of RWH (UNEP, 1999)

Advantages	Disadvantages
Providing water at the point where it is needed	Complete dependence upon the frequency and amount of rainfall
Allowing for owner operation and management	Potential shortages during dry spells or prolonged droughts
Providing a water supply buffer for use in times of emergency or breakdown of the public water supply systems, particularly during natural disasters	Contamination of water if the storage tanks are not adequately covered, if catchment areas are not properly kept clean, and if appropriate treatment is not implemented
Providing a good supplement to other water sources	Uncovered or poorly covered storage tanks can be unsafe for small children
Physical and chemical properties of the rainwater may be superior to those of groundwater or desalinated water that may have been subjected to contamination	Requires maintenance to minimize wasted water through broken gutters, drainpipes, storage tanks, and outlet taps
Construction, operation, and maintenance are not labor demanding Relieving pressures on other water sources	

2.3.5 Structural and economic aspects

The quantity of water available from a rainwater harvesting system depends on numerous factors, including the size of the catchment surface, the percentage catchment surface area that is guttered, the efficiency of the gutters in transporting the water, the size of the storage tank(s), and rainfall pattern. The demand for water must also be taken into account in order to appropriately design a system.

The costs of a rainwater harvesting system greatly vary depending on the location, type of materials used, and the level of implementation (UNEP, 1999). The majority of the costs are attributed to the roofing materials, gutters, conveyance pipes, and storage tanks. For 1995 prices, the storage tank costs surveyed by UNEP ranged from an average of 125 USD for a 1500 liter plastic tank in Saint Lucia to a 10,000 liter ferrocement tank for 1,500 USD and a 2,500 liter high density polyethylene (HDPE) tank for 500 USD in the Maldives.

2.4 SUMMARY

This chapter presented a brief background on the water aspects of sustainable buildings, primarily focusing on rainwater harvesting. The collection of rainwater from roofs can help to curb water deficits and provide an adequate water supply to those regions in need. It is important to be aware of the water quality indicators that should be tested for in order to provide clean water. These include, but are by no means limited to, TSS, turbidity, total and fecal coliforms, sulfates, color, nitrates, and nitrites. If high concentrations of these parameters are found in water, they can cause (or be an indicator of) a number of health problems, including methemoglobinemia (nitrates and nitrites) and diarrhea (fecal coliforms). There are many organizations that provide standards to regulate these parameters in drinking water, including the WHO. Microbiological and chemical contamination of RWH systems from roofs can be reduced by installing filters and first flush devices, periodically cleaning rooftop surfaces, gutters, and storage tanks, and controlling emissions of air pollutants in the surrounding area. Two key lessons were learned from the RWH examples presented in this chapter: (1) rainwater collected from roofs can be unsuitable for drinking if proper contamination prevention and treatments methods are not employed and (2) thatch roofs are not effective for RWH.

3
Background
Wastewater aspects of sustainable buildings

3.1 INTRODUCTION: WASTEWATER ASPECTS OF SUSTAINABLE BUILDINGS

This chapter presents background information on the wastewater aspects of sustainable buildings that are pertinent to the Kenya and Panama case studies. The wastewater/human waste disposal concepts presented here include septic tank systems, wastewater stabilization ponds (WSPs), phytoremediation, constructed wetlands (CWs), ecological sanitation (EcoSan), greywater and blackwater, composting, and composting eco-toilets. The advantages and limitations to each system are provided, as well as scientific and technical information regarding their function. Two examples of eco-toilet implementations are

ISBN: 9781843393238. Published by IWA Publishing, London, UK.

provided through cases in Zimbabwe and Mozambique. The waste systems such as WSPs, CWs, and composting eco-toilets, can serve as low-cost methods of dealing with excreta from buildings in an environmentally-friendly and economically beneficial manner.

3.2 WASTEWATER SYSTEMS PRESENTED IN PANAMA CASE STUDY

3.2.1 Septic tank systems

A septic tank is a watertight tank (usually underground) that receives wastewater. The tank stores the wastewater for about two to four days, where the settleable solids settle to the bottom and floatable solids such as oil and grease rise to the top of the tank. During this time, the organic solids in the tank undergo anaerobic bacterial decomposition. Septic tanks normally remove 50% of biochemical oxygen demand (BOD) and suspended solids (SS) (UNEP, 2002a). The remaining sludge produced is emptied about every three to five years (UNEP, 2002a). The tank allows for the separation of solids from liquids, digestion of organic matter, and the storage of the solids while the clarified liquid (the liquid that has been separated from the solids) undergoes further distribution and treatment (USEPA, 2005b).

The clarified liquid from the tank can be distributed in many kinds of soil absorption systems, including through the use of a leach pit (or seepage pit), seepage trench (also known as a leachfield or a drainfield), and evapotranspiration bed. A leach pit is an underground excavation sized to allow the percolation (movement of water through openings in the soil) of the volume of wastewater generated; this works well in soils with high permeability. In soils with lower permeability, a drainfield can provide a larger surface area for percolation. The drainfield is normally filled with gravel, the wastewater is piped through the gravel layer, and soil is placed above this gravel layer to the ground surface. In a drainfield, a biological clogging mat forms that contributes to an even distribution of waste in the drainfield. There should be a couple of feet of unsaturated soil beneath the drainfield before the wastewater reaches the bedrock, impervious soil layer, or water table. Thus, it may be infeasible to install septic tank systems in heavy-rainfall areas. In areas where soil permeability is very low (such as clayey soil and hard rock), an evapotranspiration bed is often used. The bed is designed much like the drainfield, only local vegetation tolerant of high nutrients and water are planted above the bed.

If properly maintained, septic tank systems can operate indefinitely. The USEPA predicts that since most systems are not well-maintained, the functioning life of septic systems is 20 years or less (USEPA, 2005b). In the United States, the installation of a septic system ranges from 1,500 USD to 8,000 USD; the costs depend on the geographical region, system size and type, and the specific soil and geological characteristics of the selected site (USEPA, 2005b). Figures 3.1, 3.2, and 3.3 show diagrams of septic tanks, a seepage pit, a seepage trench, and an evapotranspiration bed. Table 3.1 lists the advantages and disadvantages of septic tanks.

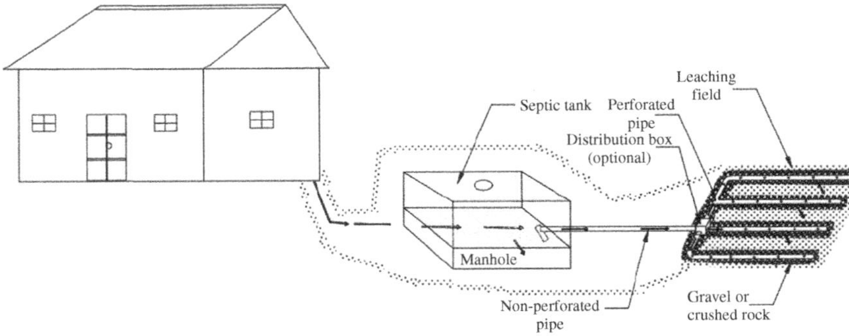

Figure 3.1 Example of a septic tank-leaching field system

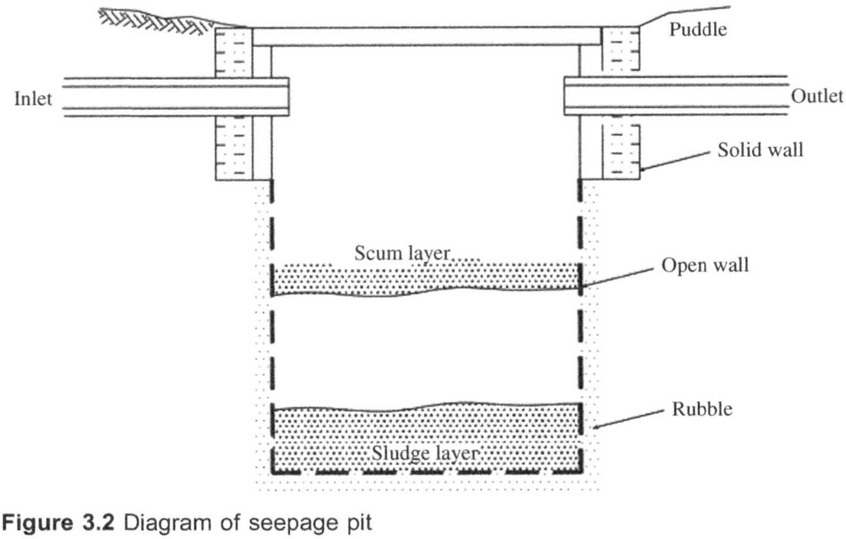

Figure 3.2 Diagram of seepage pit

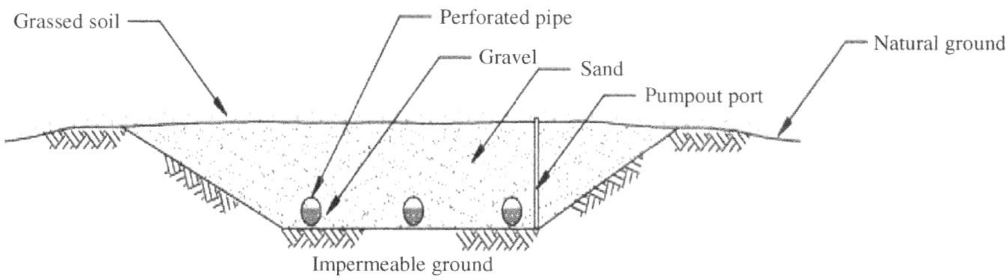

Figure 3.3 Diagram of evapotranspiration bed

Table 3.1 Advantages and disadvantages of septic tanks (USEPA, 2005b) (in comparison to other wastewater disposal systems)

Advantages	Disadvantages
Low maintenance requirements	Sludge may cause an odor problem if the sewage remains untreated for an extended period
Nutrients in waste are returned to soil	Siting limitations for septic systems include natural soil type, permeability of bedrock, groundwater elevations, and site topography
A properly designed, well-maintained system can last for more than twenty years	Improperly functioning systems can introduce nitrogen, phosphorus, organic matter, and bacterial and viral pathogens into the surrounding area and groundwater

The need for an alternative to the septic tank-drainfield system is due to the many pollution impacts that can arise. For example, odorous septic tank effluent can move about on the ground, seep into nearby stormwater drains, and thus cause health risks to people. Much of these impacts can be attributed to a poor-performing drainfield; there may be inappropriate soil conditions, excessive organic/hydraulic loading, and/or a faulty design.

3.2.2 Wastewater stabilization ponds

A Wastewater Stabilization Pond (WSP), also called a lagoon, is a man-made pond built to contain wastewater. WSPs are normally built as excavations into the ground, with a depth of one to two meters. They can be unlined, allowing for percolation of the wastewater into the soil and groundwater, or they can be lined with a layer of clay or impermeable plastic membrane to protect groundwater. WSPs have been effective in reducing biochemical oxygen demand (BOD) and suspended solids (SS) to the same levels as mechanical treatment plants, removing pathogenic bacteria and viruses better than mechanical treatment plants (due to longer residence time of wastewater in lagoon), and usually removing cysts of parasites and helminth eggs (UNEP, 2002a).

The treatment processes in the pond take place as follows. Sedimentation of the solids in the wastewater occurs once the water enters the pond. Then, aeration of the water from the atmosphere occurs by a process of diffusion. This diffusion is aided by wind movement across the surface of the water. In addition, oxygen is provided by algae in the lagoon that thrive on the nutrients (mostly nitrogen and phosphorus found in excreta). This source of oxygen is only available when there is sunlight since sunlight is needed for the photosynthetic activity of algae (UNEP, 2002a). Bacteria take in oxygen and release carbon dioxide, while algae take up carbon dioxide released by the bacteria and produce the oxygen for bacteria. Depending on the oxygen demand of the bacteria in the pond, the following conditions occur (UNEP, 2002a):

- **Anaerobic pond**: The oxygen demand of the bacteria exceeds oxygen supply by surface aeration and algal photosynthesis. Biodegradation of the organic wastes is by anaerobic bacteria. Methane gas is a by-product. Odorous gases are produced, but impact is reduced when a layer of scum forms at the water surface.
- **Facultative pond**: The oxygen demand of the bacteria is met by surface aeration and algal photosynthesis, but is not met when the latter is not active (no sunlight). The water environment is aerobic during the day, but it turns anaerobic at night. Biodegradation of organic wastes is by facultative bacteria, which can operate under both aerobic and anaerobic conditions.
- **Aerobic pond**: The oxygen demand of the bacteria is met by surface aeration and algal photosynthesis.

See Figure 3.4, Tables 3.2 and 3.3 list the advantages and disadvantages of anaerobic and aerobic ponds for diagrams on the relationship between bacteria and algae in a wastewater lagoon.

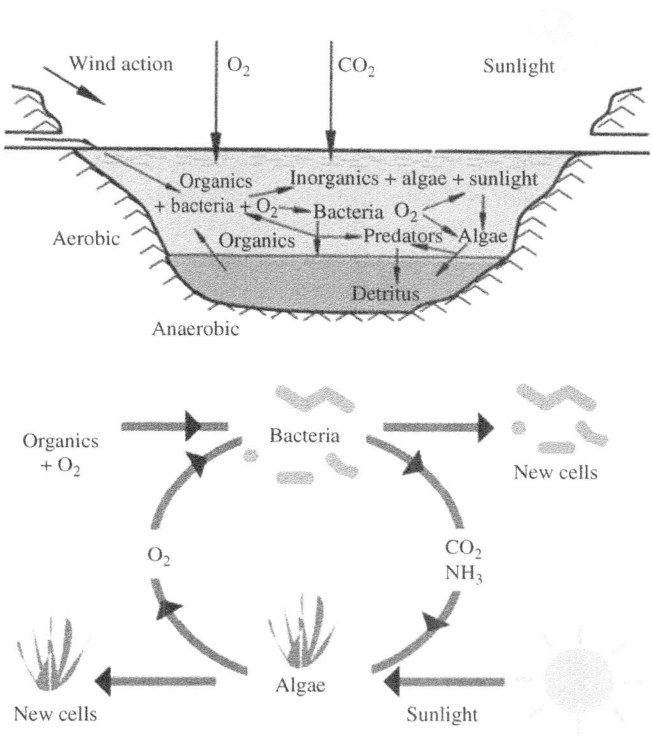

Figure 3.4 Diagrams of the relationship between bacteria and algae in a WSP (UNEP, 2002a)

3.2.3 Phytoremediation

Phytoremediation is the biological remediation of environmental problems using plants. It is the basic concept behind the use of constructed wetlands (CW) to treat waste. Table 3.4 provides the advantages and disadvantages of phytoremediation.

3.2.4 Constructed wetlands

A Constructed Wetland (CW), also known as a macrophyte trench or phytodepuration, is a wetland specifically created for the purpose of pollution control and waste management, at a location other than existing natural wetlands (Thomas & Greene, 1993). The CW serves as an alternative to a standard drainfield system in which the excreta of a septic tank flows into a soakage pit. There are two basic kinds of constructed wetlands, the free water surface wetland and the subsurface flow wetland. The subsurface flow wetland is considered to have more advantages than the free surface wetland because the flow is maintained below the media surface and there is little risk of odors, insect vectors, or public health problems. Constructed wetlands work in the following manner (See Figure 3.5). Floating and emergent plants transfer oxygen into the bed, creating aerobic microzones around the plant roots and anaerobic zones away from them. Aerobic and anaerobic bacteria become capable of breaking down the organic matter. Removal of nitrogen occurs through the denitrification process. Wetlands can greatly reduce biochemical oxygen demand (BOD), suspended solids (SS), nitrogen, pathogens, and metals through their complex chemical, biological, and vegetative uptake processes (Table 3.5).

Table 3.2 Advantages and disadvantages of anaerobic ponds (USEPA, 2005b) (in comparison to other wastewater disposal and treatment methods)

Advantages	Disadvantages
Inexpensive to construct and operate	Require a relatively large area of land
Produces methane, which can be used to heat buildings, run engines, or generate electricity	Produces undesirable odors unless provisions are made to oxidize the escaping gases
Produces little biomass, which equates to savings in sludge handling and disposal costs	Methane can cause operational problems
Effective for rapid stabilization of strong organic wastes; makes higher influent organic loading possible	Requires a relatively long detention time for organic stabilization due to the slow growth rate of the methane formers and sludge digestion
Ponds can be operated in series	Environmental conditions directly impact operations so any variance limits the ability to control the process (for example, lagoons are sensitive to temperature fluctuations)

Table 3.3 Advantages and disadvantages of facultative & aerobic ponds (USEPA, 2005b) (in comparison to other wastewater disposal and treatment methods)

Advantages	Disadvantages
Moderately effective in removing settleable solids, BOD, pathogens, fecal coliform, and ammonia	Settled sludge and inert material require periodic removal
Easy to operate	Difficult to control or predict ammonia levels in effluent
Requires little energy	Sludge accumulation will be higher in cold climates due to reduced microbial activity
The quantity of removed material will be relatively small compared to other treatment processes	Mosquitoes and similar insect vectors can be a problem if emergent vegetation is not controlled
	Requires relatively large areas of land Strong odors occur when the aerobic blanket disappearst

Table 3.4 Advantages and disadvantages of phytoremediation (USEPA, 2005b) (in comparison to other wastewater treatment methods)

Advantages	Disadvantages
Solar-energy driven cleanup technology	Time consuming process
Minimal environmental disruption and *in situ* treatment preserves topsoil	Plants that absorb toxic heavy metals or persistent chemicals may pose a risk to wildlife and contaminate the food chain
Useful for treating a broad range of environmental contaminants	It is dependant on the growing conditions required by the plant (such as climate and temperature)
Inexpensive compared to conventional physicochemical methods	It works best when the contamination is within reach of the plant roots
	A suitable control plan is needed to keep various animals away from applications

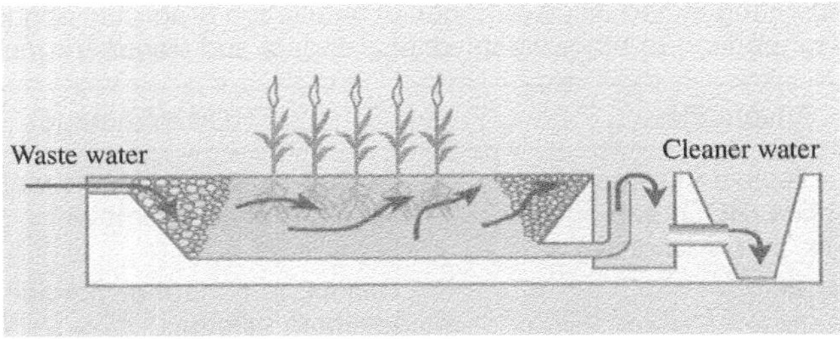

Figure 3.5 Constructed Wetlands are a technology for treating stormwater or wastewater. A constructed wetland consists of a gravel bed on which suitable wetland plants are grown. As water passes through the substrate, it is purified through the activity of bacteria attached to the gravel, plant roots, soil and other particles (UNEP, 2003)

Table 3.5 Advantages and disadvantages of constructed wetlands (USEPA, 2005b) (in comparison to other wastewater treatment methods)

Advantages	Disadvantages
Economically feasible (where septic tank systems are economically feasible)	Potential clogging problems, which can be prevented through effluent sand filters to reduce loading
Can be combined with other technologies such as sand filters (hybrid systems)	Caution must be exercised in malaria regions because the constructed wetlands can attract mosquitoes
Decrease harmful impacts of traditional septic tank-drainfield systems	A suitable control plan is needed to keep various animals away from applications

3.3 WASTEWATER SYSTEMS PRESENTED IN KENYA CASE STUDY

3.3.1 Ecological sanitation (EcoSan)

Ecological sanitation (EcoSan) is sanitation whose design builds on the concept of protecting ecosystems, and which treats excreta (feces and urine) as a valuable resource to be recycled. EcoSan seeks to make minimal use of freshwater, while also minimizing the discharge of waste into the ecosystem.

Studies show that "each person's waste can provide enough nitrogen, phosphorus, and potassium to grow a year's supply of wheat and maize for that person" (Andersson, 2005). Thus, EcoSan can help poverty reduction by supporting the livelihoods of people, and help to create better lives for women and children. Ecological sanitation can be viewed as a three-step process dealing with human excreta: containment, sanitization and recycling. EcoSan is based on the following principles (UNDP, 2005):

- **Disease Prevention**: must be capable of destroying or isolating fecal pathogens
- **Environmental Protection**: must prevent pollution and conserve valuable water resources
- **Nutrient Return**: must return plant nutrients to the soil
- **Culturally Acceptable**: must be aesthetically inoffensive and consistent with cultural and social values
- **Reliable**: must be easy to construct and robust enough to be easily maintained in a local context
- **Convenient**: must meet the needs of all household members considering gender, age and social status
- **Affordable**: must be accessible to all households in the community

Women can also benefit directly from EcoSan. Women, who are generally involved in food crop production and concerned about food security, would be positively impacted from increase access to soil nutrients needed for crop production, as well as the possibility for small vegetable gardens and fruit trees close to homes (Hannan and Andersson, 2002). EcoSan systems involve two ways to make human excreta harmless: dehydration and decomposition. The preferred method will depend on climate, groundwater tables, amount of space, and intended purpose for the sanitized excreta.

3.3.1.1 Greywater and blackwater

Blackwater is the effluent from the toilet; it is water containing excreta. Greywater, also known as sullage, is the effluent from all other sources including a washbasin, laundry, bathroom, and kitchen (water that has been used for washing, cooking, etc.). Greywater generally contains less pathogens than blackwater, has far less nitrogen than blackwater, and decomposes faster than blackwater. It can be treated and re-used for toilet-flushing, clothes washing, and garden watering. Blackwater, since it is rich in nitrogen and phosphorus, can serve as an ideal fertilizer.

3.3.1.2 Composting

The key behind most EcoSan implementations is composting. Composting is the controlled biological decomposition of organic matter into a humus or soil-like material. Composting is an effective procedure of treating wastes with high organic content, especially blackwater. Composting is an accelerated biooxidation of organic matter passing through a thermophilic stage where microorganisms liberate heat, carbon dioxide and water. Composting can produce natural fertilizer for home-grown crops, community gardens, and other agricultural uses. Composting methods are appropriate means for EcoSan, as they can be used as an eco-toilet system (Table 3.6).

Table 3.6 Advantages and disadvantages of composting (in comparison to other wastewater disposal and treatment methods)

Advantages	Disadvantages
Reduces populations of pathogenic microorganisms	The nutrient composition of the compost varies a great deal
Low, medium and high technology systems are available	Slow rate of decomposition
Reduces waste bulk density	
Compost does not cause excessive weed growth, as is the case with ordinary farm manure	
Good crops can be obtained without the need for extra chemical inputs	
Inexpensive	

3.3.1.3 Vermicomposting

Vermicomposting is composting that involves the joint action of earthworms and microorganisms. The earthworms are responsible for turning, fragmentation and aeration. The waste materials and microbes pass through the worm's body, where its digestive system converts the substrate into worm excreta. Worm excreta, or vermicastings, are usually a stable material enriched with plant nutrients and soil microbes (Panikkar, 2003). Only very strong soil microbes, which come from the indigenous microbial community in the waste matrix, are able to escape the digestive enzymes of worms.

One example of successful vermicomposting can be found in India. Rajesh Patil, a farmer of Jalke village in India has completely shifted to organic farming, using the end products of vermicomposting as a micronutrient. By doing this, he saves approximately (Indian Rupees) 60–70 thousand Rs per year, which were spent earlier on chemical fertilizers and pesticides. He has made approximately 20 cemented structures for vermicomposting and has noted an improvement in soil quality (IDE, 2004). Patil cultivates turmeric, and since it is not very common in the region, it commands a good price. He plans to work specifically with small farmers to train them to grow turmeric as a high value crop (Table 3.7).

Table 3.7 Advantages and disadvantages of vermicomposting (in comparison to other wastewater disposal and treatment methods)

Advantages	Disadvantages
Reduces populations of pathogenic microorganisms	Nutrient composition of the compost varies a great deal
Low, medium and high technology systems are available	Slow rate of decomposition
Compost does not cause excessive weed growth, as is the case with ordinary farm manure	Vermicomposting requires more management and maintenance than other composting systems to maintain healthy worms
Good crops can be obtained without the need for extra chemical inputs	
Inexpensive	
Supplies a suitable mineral balance, improves nutrient availability and could act as complex-fertilizer granules	

3.3.2 Eco-toilets

3.3.2.1 Dry, double-vault urine-diversion toilet

One EcoSan composting system is the dry, double-vault, urine-diversion eco-toilet. These types of toilets are being used in various countries, including China, India, Vietnam and Mexico. The toilet is designed to separate urine from feces (in two separate vaults) so that they can be used as fertilizer. The urine can be used immediately as fertilizer (usually diluted with water), while the feces must be composted. Materials like wood ash, lime and soil, are added to cover excreta for drying purposes. Ash and lime increase pH, which helps to destroy pathogens. It takes an average family six months to fill one of the vaults. After the first vault is filled, it is emptied following an additional 6 months of sanitization and the material is taken to soil compost. The toilet is not odorous, does not attract flies and is an affordable EcoSan solution.

3.3.2.2 Example: Urine diversion system in South Africa

Richard Holden, Mvula Trust's national sanitation operations manager, has installed a dry eco-toilet system into an urban home in Johannesburg, South Africa. In his home, Holden uses a simple configuration of urine diversion. The urine is led into a container, and the feces fall into the pit below. A mixture of soil and ash is added to the pit to help kill off the pathogens and dehydrate the feces. If made by oneself, the cost is 30 R (South African Rand), or about 5 USD (United States Dollar). According to Holden, even if the toilet is made commercially, "the production is so simple it can be made within a community, promoting local economic development" (Holden, 2002). The urine from the toilet is used as a fertilizer. It is removed once a week and poured into a compost pile around the garden. Every six months the dried feces are removed from the back of the toilet and also poured into the compost heap. The Mvula Trust has introduced about 2,500 urine diversion toilets in South Africa.

3.3.2.3 Fossa Alterna (FA-type) and Arbour Loo (AL-type)

Toilets such as the *Fossa Alterna* and *Arbour Loo* have been successfully used in many countries, including Mozambique and Zimbabwe.

The FA-type eco-toilet uses two alternating pits with a similar frequency of alternation as a double-vault dry eco-toilet. Once sanitized and composted, the contents are removed and used in agriculture. The Fossa Alterna can be used continuously on the same site by alternating between one pit and the second, with the humus being excavated once a year. This allows for an empty pit every year.

The AL-type eco-toilet is a single shallow pit which receives soil additions after each use and a tree is planted in the pit when it is full. After a period which can vary from a few months to a few years, depending on the pit size and extent of use, the Arbour Loo eco-pit will be filled with a composting mix of ingredients. The toilet itself moves to a new chosen site, where the process is repeated.

3.3.2.4 Example from Mozambique

A successful EcoSan program exists in the town of Lichinga in the Niassa Province in Mozambique. Lichinga, hosting about 85,000 inhabitants, represents a typical small town in the developing world. It has poor infrastructure, and is unable to meet the needs of local residents. Families live close together, water and sanitation problems are acute, economies are weak, and people are heavily dependent on agriculture. In this situation, eco-toilets as opposed to flush-toilets are an appropriate option. Two types of eco-toilet systems were built in Lichinga: FA-type and AL-type. Both of these types do not have flies or odors, as do traditional pit toilets in Niassa Province (Winbald, 2004). In both types, dry soil is added to the toilet in at least an amount equal to the volume of the solid excreta just deposited. The soil absorbs, purifies, and converts the excreta into safe, rich, and nutrient humus. Trees currently being tested in AL-type pits in Niassa include guava, mango, orange and avocado as well as a range of local fruit trees. The advantage of the FA-type toilet is that there is no need to dig new deep pits. The total cost of an FA-type toilet in Niassa Province is about 25 USD (Winbald, 2004). See Figures 3.6 and 3.7 for diagrams of the FA-type and AL-type toilet systems.

3.3.2.5 Example from Zimbabwe

The non-governmental organization (NGO) Mvurumanzi Trust of Zimbabwe has instituted about 1,847 eco-toilets in the country. There is mass adoption of Arbour Loos and Fossa Alterna type toilets in peri-urban areas. Unfortunately, there were problems with the knowledge transfer of how to correctly operate the eco-toilets. Those who were reusing the urine had not fully understood how it should be diluted before use. Some failed to mix the urine with water and this resulted in the loss of vegetables gardens.

At first people did not accept the notion of having urine diversion system toilets in their yards as it is against many cultures and traditions to handle feces. Consequently, Mvuramanzi ran community workshops to tell people why urine diversion toilets could be beneficial. Thomas, the son of a headman whose family is using a urine diversion toilet, said: "We have realized that the old Blair latrine smells and have lots of flies, but the UDS [urine diversion toilet] does not have these problems" (Guzha and Musara, 2003).

In Harare, an assessment of the social impacts of eco-toilets was completed. In a survey, only 6.7% of the people felt that it was easy to maintain the toilets (Guzha and Musara, 2003). 76.5% of respondents said that they like eco-toilets because they are convenient and private to the household The "health and hygiene benefits became incidental" (Guzha and Musara, 2003). Thus, the Harare case study illustrates that social reasons, in addition to environmental reasons, can promote the use of eco-toilets. It also stresses the importance of maintenance.

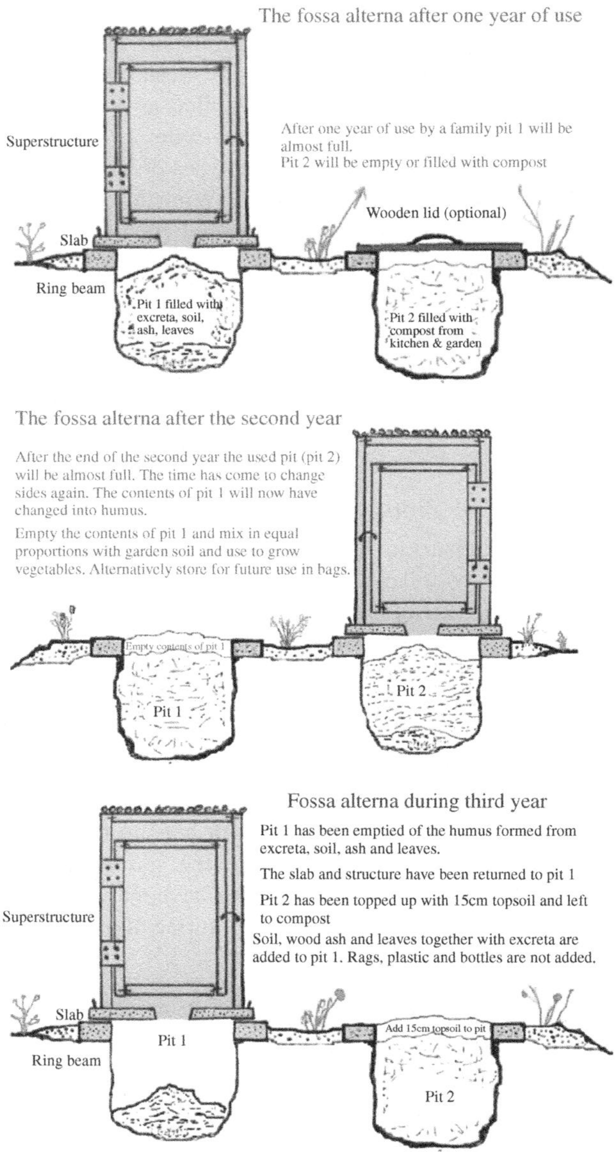

Figure 3.6 Different stages of FA-type eco-toilet (Morgan, 2002). Figure from *an ecological approach to sanitation in Africa: a compilation of experiences* by Peter Morgan, 2004. Courtesy of Stockholm Environment Institute

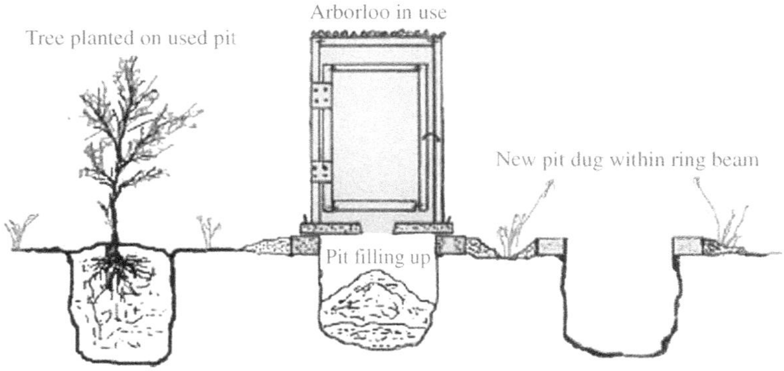

Figure 3.7 Different stages in AL-type eco-toilet (Morgan, 2002). Figure from *an ecological approach to sanitation in Africa: A compilation of experiences* by Peter Morgan, 2004. Courtesy of Stockholm Environment Institute

3.4 SUMMARY

This chapter provided a brief background on the wastewater aspects of sustainable buildings that are relevant to the Kenya and Panama case studies. The septic tank system can be simple and reliable, but can cause contamination to the soil if malfunction occurs. Two kinds of WSPs, anaerobic and aerobic, are relatively inexpensive and easy to operate, but they require large areas of land and can create odors. The use of plants to degrade pollutants is another low cost method of treating wastewater. Constructed wetlands may attract animals including mosquitoes as do WSPs. Ecological Sanitation is a concept in which excreta is treated as a valuable resource. Composting excreta for use as fertilizer can help to create good crops to be sold or used for food security; however there is a slow rate of decomposition and the composition of the compost can vary (and thus must be monitored). Rajesh Patil, a farmer in India, was able to save 60–70 thousand Rs (that would be spent on chemical fertilizers and pesticides) each year by using the end products of vermicomposting as a micronutrient for his crops. The main types of eco-toilets in use in developing countries are the urine diversion, Fossa Alterna (FA), and Arbour Loo (AL) types. Two key lessons were learned from the Zimbabwe example in which 1,847 eco-toilets were instituted in the country, and from the Mozambique example in which FA and LA type of eco-toilets were installed: (1) crops such as mango, orange, guava, and avocado can grow effectively in AL-type pits and (2) cultural barriers can exist in regards to working with human excreta.

4
Panama Site and Water Management Practices
The Smithsonian Tropical Research Institute sustainable building in Bocas del Toro

4.1 INTRODUCTION: PANAMA SITE AND WATER MANAGEMENT PRACTICES

This chapter presents information regarding the water aspects of the STRI laboratory building in Bocas del Toro, Panama, as well as relevant water issues in the surrounding community. The following are included: a description of the potable water and wastewater systems in Bocas del Toro, the motivation for

ISBN: 9781843393238. Published by IWA Publishing, London, UK.

constructing a sustainable laboratory building at STRI, and water management at STRI that includes rainwater collection, potable water treatment, wastewater treatment, and water conservation methods. The data collected while interviewing officials in Panama indicate that wastewater disposal and municipal potable water in Bocas del Toro is inadequate. The STRI building aims to provide a proper wastewater treatment and disposal method through a wastewater stabilization pond (WSP) and constructed wetland (CW) system; it also aims to provide good water quality through a rainwater harvesting system. In order to determine (1) the suitability of the harvested rainwater for drinking purposes and (2) what kind of treatment (if any) is needed, water quality experiments were completed that tested the rainwater for total coliform, fecal coliform, sulfates, nitrates, nitrites, pH, true and apparent color, total suspended solids (TSS), and turbidity. Chapter 5, Water Quality Testing at Panama Site, gives a description of the types of tests that were conducted and presents the findings.

4.2 DESCRIPTION OF STUDY AREA AND STRI BUILDING

The Panama study was completed at the Smithsonian Tropical Research Institute (STRI) building in the province of Bocas del Toro, on the island of "Isla Colón," and in the town of Bocas del Toro, Panama. Panama is located in Central America, with bordering countries Costa Rica to the west and Columbia to the east. Bocas del Toro Province is bordered by Costa Rica to the west, Chiriquí Province to the south, Veraguas Province to the east, and the Caribbean Sea to the north. See Figure 4.1 for a map of Panama.

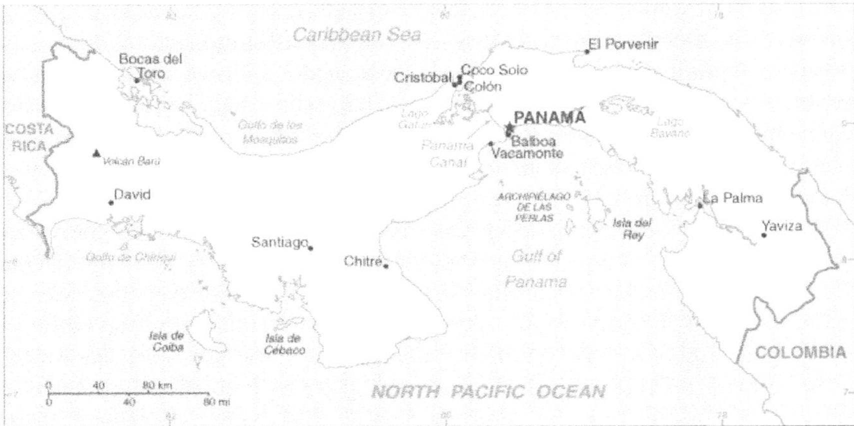

Figure 4.1 Map of Panama (from University of Texas Libraries)

The STRI building is located at a former sawmill, on a sensitive coastal site next to a mangrove swamp. The Province of Bocas del Toro has a territorial extension of 8,917 square kilometers (about 12% of the national territory), while the town of Bocas del Toro is about 2,133 square kilometers (about 2.9% of the total area of the country) (STRI, 2000). The population of the Bocas Province in 1999 was projected at 136,604 inhabitants, with an annual growth rate of about 3.5% (STRI, 2000). The main activities of the province are the growing of banana for export and tourism.

Bocas has a very humid tropical climate, with average temperatures of 31.5°C in the summer and 25.6°C in the winter, and an average relative humidity of 80 to 85 percent (Pascal, 2000). The Bocas area receives abundant rainfall. The annual average rainfall for Isla Colón is 3,000 mm, or 118 inches (Pascal, 2000). Although it does not have strict rainy and dry seasons, the least rainy times of year are mid-August to mid-October and February to March when the seas are relatively calm.

The Smithsonian Tropical Research Institute conducts research on the ecology, behavior, and evolution of tropical organisms. The new STRI building was constructed to provide needed facilities for their research. The building includes labs for resident and visiting scientists, teaching labs, a conference room, and support spaces. The main laboratory building is designed to be a state-of-the-art scientific facility that minimizes its environmental impact. The guiding principle for the design was that the building should not have any negative impact on the environment; the design allows for the collection of its own water, the treatment of its own waste, and the generation of its own energy.

The STRI building encompasses 7,530 square feet (700 square meters) and serves as a research facility for scientists. Most of the construction was completed by October 2003; construction completed by this date is known as phase 1. All construction thereafter is known as phase 2. Three water aspects of the building are examined: the rainwater harvesting and treatment system (completed in January 2006), the wastewater stabilization pond and constructed wetland system (to be built at an un-specified later date), and water conservation methods employed in the building.

The total cost of the building for phase 1 (does not include the three sustainable aspects considered in this study) is 1.3 million USD (USDOE, 2004). It was funded exclusively through donations to the Smithsonian Institution. Apart from the water aspects of sustainable buildings, the building incorporates a photovoltaic roof that provides electricity, uses local hardwood as the primary building material, and overall seeks to limit the environmental degradation often brought by new construction. Six "green strategies" were employed (or plan to be employed) by STRI, namely, development impacts, runoff reduction, rainwater collection, integration with site resources, siting analysis, and low-impact siting (USDOE, 2004). Development impacts included restricting vehicle access during construction to reduce damage to vegetation and limiting the parking area. Runoff reduction consisted of reducing the driveway pavement. Rainwater collection was achieved by collecting and storing rainwater for uses in the building. Integration with site resources will be achieved by providing a constructed wetland for wastewater treatment. Siting analysis included assessing regional climatic conditions, creating a map of the physical elements on the site (structures, topography, soils, hydrology), and creating a map of vegetation on site, including notation of significant specimens. Low-impact siting included locating the building(s) on the most degraded part of the site and siting development carefully to protect significant ecosystems.

One of the key lessons learned from constructing this building was the difficulties that can arise from constructing a sustainable building of this magnitude and for this purpose (laboratory facility primarily for visiting scientists) in a remote location. The project was initially on a fast schedule in which construction began before drawings were complete. However, when full pricing came in, some features of the building (such as the rainwater harvesting system and wastewater treatment system) had to be cut or delayed until phase 2, even though it would have been more efficient to do them in phase 1 (USDOE, 2004). The estimated cost of the RWH system is 13,750 USD, the rainwater treatment system cost approximately 17,350 USD, and the current septic tank system cost roughly 3,750 USD.

According to Sheryl Kolasinski of STRI, there must be an underlying philosophy for both the science and architecture of the Smithsonian Institution. She stated, "Fundamentally, we care about the environment. We love and study the environment, so we do not want to ignore that when building facilities" (personal communication, February 27, 2006). The motivation for building a green building for the Bocas del Toro site emanated from that philosophy. With the direction of the deputy director of STRI at the time, Tony Coates, the Caisa firm performed a preliminary environmental impact study of the Bocas site. The term "sustainability" was not a part of the vocabulary at that time. The approach was aimed at answering the questions, "How does one treat this subject and what is the best way to deal with the development" (personal communication, February 27, 2006). The members of the planning team recognized that the site was special; it was where a precious mangrove, swamp, and coral reefs came together. In addition, decisions were made to make certain that a highly advanced scientific research center could exist while maintaining some of the traditional architecture of the region. At the end of the day, the building had to be able to meet the conditions of the Tropics and the needs of the researchers. It had to be built to withstand high temperatures and humidity, there had to be uninterrupted electricity for the freezing of samples and computer downloads, and a variety of other conditions had to be met relating to the environment.

4.3 POTABLE WATER AND WASTEWATER IN BOCAS DEL TORO

4.3.1 Municipal potable water in Bocas del Toro

The source of potable water for the town of Bocas is a man-made freshwater lake called Big Creek. The volume of Big Creek is mostly made up of rainwater. The water is pumped from the lake up-hill to a water purification plant. There, the water undergoes both physical and chemical steps of water purification. These include screening, flocculation, rapid sand filtration, and finally disinfection.

First, the water is filtered through a sieve to remove large objects such as leaves and insects. Then, the water is treated with small volumes of appropriate chemicals which will form a chemical floc (clump of impurities) that entraps particles. The water undergoes six main steps of flocculation: the addition of aluminum sulfate, hydration with white lime, the addition of fluorosilicate, the addition of an anionic polymer, and the addition of a cationic polymer. Then, the flocculated water is passed through a rapid sand filter that strains out the floc and the particles trapped within it. After a certain period of time, the sand filter becomes clogged with floc, and the water is then backwashed to remove the floc. The backwash water is pumped into settling tanks so that the floc can precipitate out and is then disposed of as waste material. The plant uses sand filters that combine small stones and sand. A pile of black sand about ten inches (25.4 cm) in height lays on top of a pile of white sand that ranges from two to four feet (0.61 m to 1.22 m) in height. The filter is changed every two years, and one part of the filtration system is cleaned every sixty days.

The final step in the purification process is the addition of chlorine as a disinfectant. Chlorine is a common and effective disinfectant which destroys disease-causing organisms in water. In order to chlorinate the water properly, the right amount of chlorine should be added. Chlorine demand is the amount of chlorine required to kill bacteria, oxidize iron or other elements in the water, and oxidize any organic matter that may be present. The chlorinated water is tested for free available chlorine residual. If the chlorine demand is greater than the amount of chlorine introduced, there will be no free available chlorine residual and thus more chlorine should be added. When the chlorine gas pump is not working at the plant, the plant uses chlorine tablets. The chlorinated water is then pumped into storage tanks; from there, the water is transported through pipes to the community.

The water purification process used by the town of Bocas del Toro is standard for Panama. One problem is that the plant was built for a much smaller water demand than currently exists. According to Juan Villagra from the Institute of Aqueducts and National Sewage Systems (IDAAN) that control the water treatment plant in Bocas, the plans for the plant were set at a water demand of *300 gallons per day (1,140 liters per day)*, but the government built the plant even smaller than that (Villagra, 2005). Thus, the exact capacity of the plant is unknown. However, the plant is currently servicing a water demand of *40,000 to 50,000 gallons per day (151,400 to 189,300 liters per day)*. Thus, anytime production increases, the plant operators from IDAAN must send the water through the purification system so fast that it does not have time to properly settle. Therefore, sometimes there are sediment particles in the community potable water, rendering it unsafe to drink. In addition, there are technical flaws in the water treatment process; water pressure is very irregular (STRI, 2000).

The Ministry of Health lists diarrhea, intestinal problems, and parasithosis as the highest causes of infant mortality in Bocas del Toro (STRI, 2000). These health problems are directly linked to drinking water contamination. One can deduce that if the water quality was better in Bocas, the infant mortality rates would decrease. There is a two million dollar upgrading program underway to extend the size of the plant to accommodate the current water demand. Figures 4.2 to 4.4 show Big Creek, the water treatment plant, and one of the water storage tanks used for the Bocas municipal water.

4.3.2 Municipal wastewater treatment in Bocas del Toro

The town of Bocas del Toro uses an aeration pond to treat its wastewater. However, only 36% of the population of Bocas is connected to this system. Approximately 47.7% of the population disposes of their excreta in septic tanks, of which 32.8% are in poor condition or inadequately constructed (STRI, 2000). A great quantity of latrines in the Bocas town spill directly into the sea. The aeration pond is in very poor condition; it has odors that reach into the nearby community. A major cause of this condition is various holes in the impermeable buffer that lines the pond. The buffer is meant to separate the pond from the rest of the land. Because the rubber lining is breaking apart, the sewage in the pond leaks underground to locations outside of its boundaries. One can see and smell sludge leading up to the wastewater pond.

Figure 4.2 Big Creek, source of municipal water for Bocas del Toro

There is currently no filtration or purification system for the wastewater in addition to the aeration pond. Gustavo A. Baker, IDAAN Operations Manager for the municipal wastewater stabilization pond in Bocas del Toro, stated that IDAAN added chemicals to treat the wastewater in the pond "very many years ago" (Baker, 2005). There are two tanks that take in wastewater from the pond and distribute it straight into the sea. At one time, there were pipes that connected the water from different parts of the lake to the holes in those tanks, but the pipes have been disconnected. The Panamanian Institute of Aqueducts and National Sewage Systems plans to reconnect the pipes once they do a thorough maintenance operation. The maintenance will involve drying out the pond and relining the two sides of the pond. However, the pond will not be completely dried so that aquatic species can survive. In addition, IDAAN plans to make a new pond (50 meters by 50 meters) adjacent to the existing one. Figures 4.5 to 4.8 show photographs of the aeration pond, the drum that leads the wastewater to the sea, the broken lining, and the sewage from the pond leaking to other parts of the land.

Figure 4.3 Water purification plant for potable water in Bocas del Toro

4.4 WATER MANAGEMENT AT STRI

4.4.1 Rainwater collection

4.4.1.1 Situation before January 2006

The STRI station used the same municipal potable water as the town of Bocas del Toro before January 2006. However, since the "potable" water for the town is considered unsafe to drink, STRI installed bottled water fountains. People still used the municipal water for all other purposes such as bathing, cooking (recommended that water be boiled first), flushing toilets.

Figure 4.4 Water storage tank for municipal water in Bocas del Toro

Figure 4.5 Wastewater aeration pond for Bocas del Toro

Figure 4.6 Drum that leads wastewater directly to sea

Figures 4.7 Holes in the aeration pond buffer

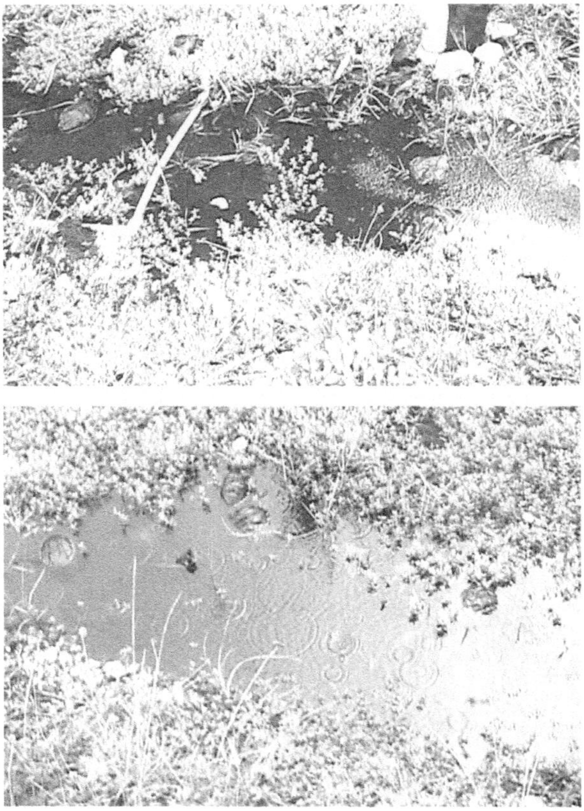

Figures 4.8 Sewage from aeration pond traveling to nearby community land

4.4.1.2 Rainwater harvesting system

Although the rainwater in Bocas is generally of good quality, when it is collected from roof surfaces it is affected by bird droppings, dust, and other substances. These substances collect on the roof during rainwater collection and are washed into the storage reservoirs. In the utilities plan for the laboratory building, it was estimated that 50 people would require water from the rainwater system (for purposes such as drinking, cleaning the area, washing of equipment, and sanitary installations); the demand for water was considered to be 50 liters per person per day, or 2.5 cubic meters per day for everyone (about 910 cubic meters per year). In order to calculate the volume of rainwater available at the site, the following equations were used:

$$\textit{Gross water supply (volume)} = (\textit{surface area of roof}) \times (\textit{rainfall}) \tag{4.1}$$

$$\textit{Net water supply (volume)} = \mathrm{C_R} \times (\textit{gross water supply}) \tag{4.2}$$

$$\textit{Net water supply/number of days in a year} = \textit{daily water storage volume} \tag{4.3}$$

$$1\,m^3 = 1000\,liters \tag{4.4}$$

where,

C_R is the roof material runoff coefficient
Gross water supply (yearly) is in m^3
Surface area is in m^2
Rainfall (yearly) is in m

The total roof area was estimated at 700 square meters with a yearly rainfall of 2,500 millimeters (2.5 meters) and with three days a week of rain. The roof area was not adjusted for pitch; the horizontal projection of the roof area was used. Since the roof pitch is fairly shallow (about 17°), there is not much difference when taking into account pitch. The gross annual supply was estimated to be 1750 m^3 (700 $m^2 \times 2.5$ m). Since the runoff coefficient for a photovoltaic roof was estimated to be about 2/3 or 0.67, the net supply was estimated at 1,100 m^3 (or 3 m^3 per day). The planners recommended that the system be designed for a net water storage of 6 m^3 per day because of the frequency of the rain, as well as an extra cubic meter of storage to allow for filtration time in sand filters used to treat the rainwater.

The actual rainwater harvesting system in place differs from initial plans. There are currently three tanks installed that each have a capacity of 1,000 gallons (3.8 m^3), although there are connections and adequate space in place for an additional 5 tanks of the same storage capacity. Thus, the builders constructed the rainwater harvesting system to have a total of 8,000 gallons (about 30 m^3) of water storage per day. Figure 4.9 shows a schematic diagram of the current water storage tanks in place.

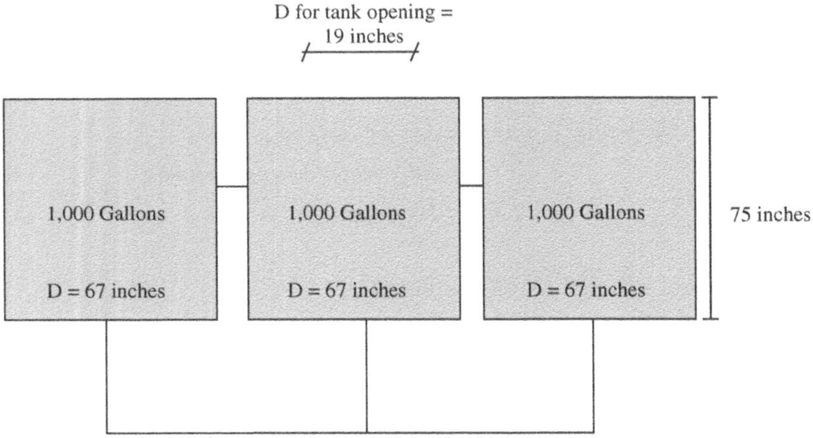

Figure 4.9 Three rainwater storage tanks in place at STRI Bocas Laboratory (Note: D stands for diameter; tanks are cylindrical and identical; 19 in ≈ 48 cm; 75 in ≈ 191 cm; 67 in ≈ 170 cm; 1,000 gallons ≈ 3,790 L)

The system works as follows. Water is collected from the STRI laboratory building photovoltaic roof. The rain from the gutter, made of aluminum, is then directed into a pipe that leads to two places: an overflow pipe system and the three storage tanks. See Figures 4.10 for photographs of the rainwater collection pipe from the roof gutter. The rainwater from the pipe is directed into the three reservoir tanks by various pipes. There are valves on each reservoir tank which can be closed to allow for tank cleaning, without affecting the rest of the system operations. The tanks are manually cleaned, without the use of a backwash system. There are two overflow mechanisms in the RWH system. In the situation in which any of the reservoir inlet valves are closed, the rainwater is directed into an overflow system that directs the overflow into a pond on the site (Figures 4.11). Thus, the valves on the tanks can be closed to allow for manual cleaning, in which case the water that would be directed into that tank is directed into the first overflow system. Figure 4.12 provides an image of the storage tanks. In the case of very heavy storms in which the amount of water coming into the tanks far exceeds the tank storage capacity, there is a second overflow system that conducts excess water from a pipe into the pond on site (Figure 4.13). Figures 4.14a and 4.14b show a schematic diagram of the complete RWH system at STRI.

Figure 4.10 Rainwater collection from roof gutter

Figure 4.11 First overflow valve and its pipe leading to pond

Figure 4.12 Storage tanks at STRI

Figure 4.13 Second overflow system for storage tanks and pipe leading to pond (pipe in foreground of picture)

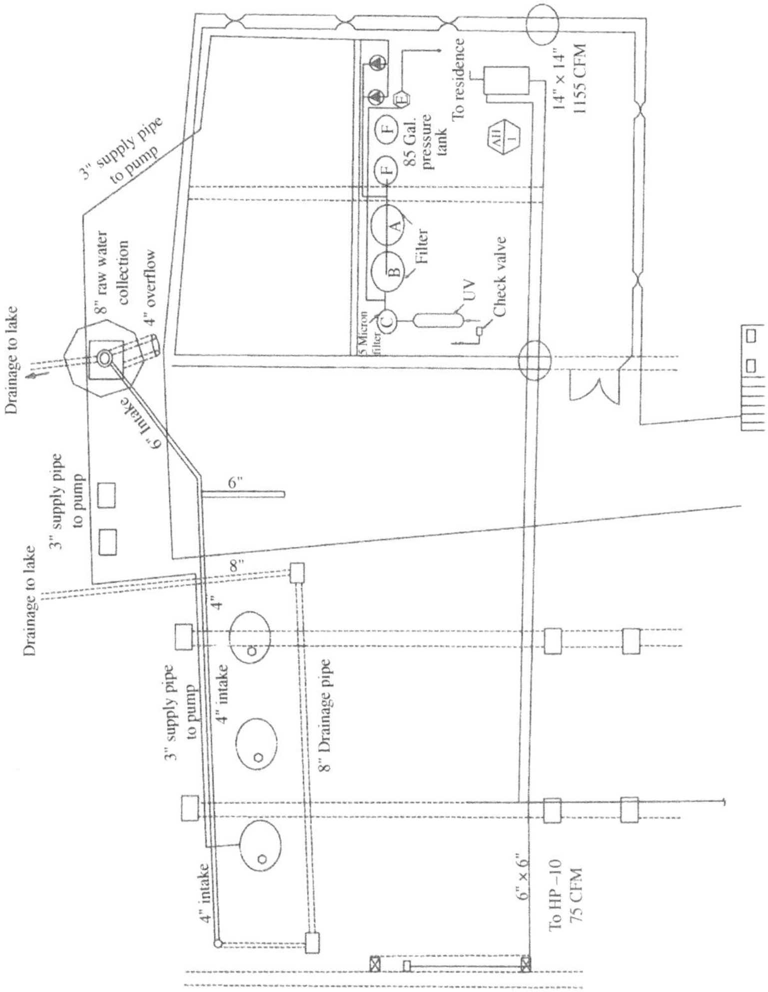

Figure 4.14a Schematic diagram of entire rainwater harvesting system (by Dennis Allen of STRI)

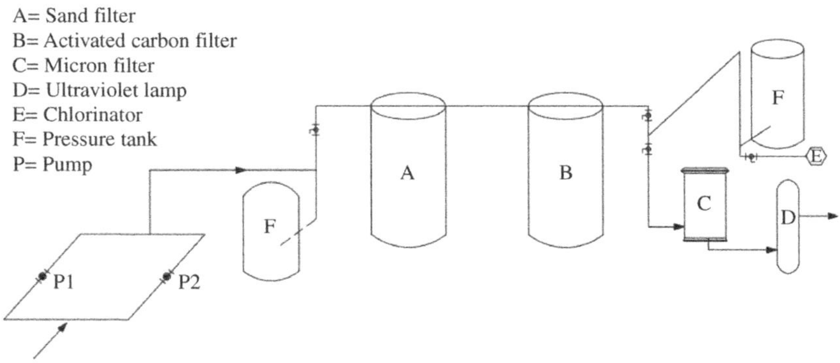

Figure 4.14b Schematic diagram of entire rainwater harvesting system (by Dennis Allen of STRI)

4.4.2 Potable water treatment

The current water treatment system for the rainwater involves more than just sand filters. The rainwater is pumped into an indoor treatment room (Figure 4.15). Figure 4.16 shows a diagram of each step of the rainwater purification system including the costs of each purification mechanism. For details on the materials used in the UV disinfection system see Figure A.1 in the Appendix. Note that there is no chlorination of the water directed to the laboratory since it will be used immediately. However, liquid chlorine is added to one 400-gallon (1,500 L) storage tank by the dormitories where the water will not be used instantly. The water purification system contains a pipe that leads from the municipal water source to the water treatment room; this allows for emergency treatment of the municipal water in the case of any malfunction or deficit of rainwater (Figure 4.17).

The treatment system is broken into five stages as follows: (A) sand filtration, (B) active carbon (AC) filtration, (C) micron filtration, (D) ultraviolet (UV) disinfection, and (E) chlorination. The sand filter combines both fine and coarse sand to filter out particulate matter (Figure 4.18). The carbon filter is most effective in removing organic contaminants from water (Figure 4.19). Because organic chemicals are often responsible for taste, odor, and color problems, AC filtration can generally be used to improve aesthetically unpleasing water. Active Carbon filtration works by attracting and storing chemicals as water passes through it. Active Carbon is made of tiny clusters of carbon atoms stacked upon one another. The carbon becomes activated by pumping oxidizing gases through the material at extremely high temperatures. The activation process produces pores that result in high adsorptive properties in the carbon filter (North Dakota State University, 1992) (Figure 4.19). The 5-Micron filter contains wound polypropylene cartridge that is 20 inches (50.8 cm); (a micron is equal to 0.000039 inches or 0.0001 cm) long (Figure 4.20). The micron filter essentially filters out particles at the micron level (a micron is equal to 0.000039 inches). Ultraviolet purification is a rapid and effective method of disinfection that does not require chemicals or heat. This system uses a Sanitron™ Ultraviolet Purifier, which utilizes germicidal ultraviolet lamps that produce short wave radiation lethal to microorganisms present in water (see Figure 4.21 and 4.22 for a photograph of the UV radiator and a detailed process description). The UV radiator used in this system operates at an apparent power of 35 kilo-volt amperes (kVA).

Figure 4.15 Pipes in indoor water treatment room

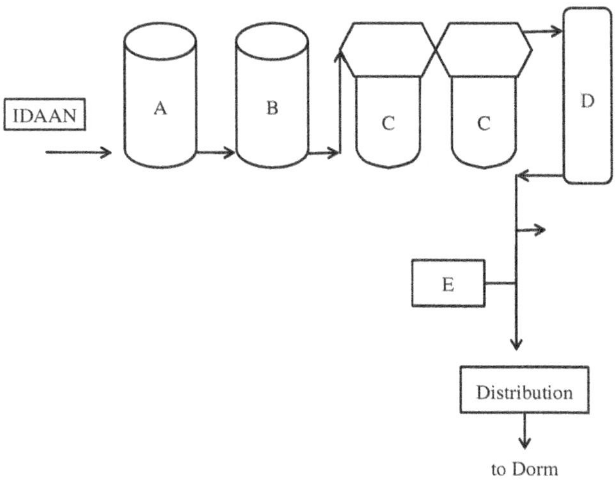

A. Sand filter
B. Activated carbon filter
C. 5 Micron filter
D. Ultraviolet disinfection system
E. Chlorination system

Sand filter	3,083.26
Activated carbon filter	3,097.20
5 Micron filter	324.76
Ultraviolet disinfection system	2,371.50
Chlorine dosage system	285.32
Tubing, adaptors and PVC pieces for plumbing	128.95
Electrical materials	83.81
Installation (labor)	450.00
Total Net Cost of properly installed system	10,416.04

Figure 4.16 Flow chart of rainwater purification system source (Based on original diagram in Spanish from IDAAN, costs are in USD)

Figure 4.17 Valves for municipal water treatment

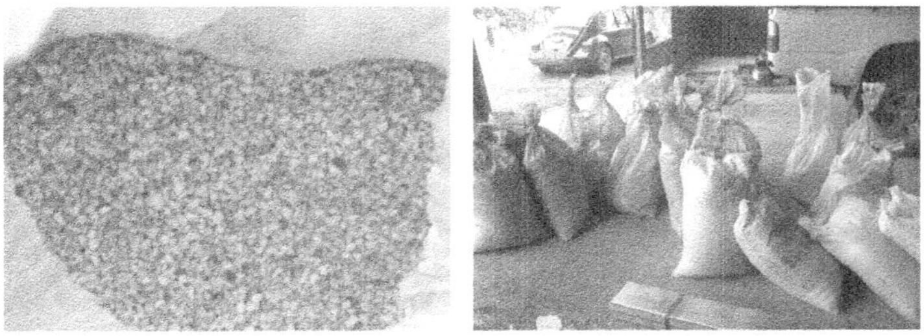

Figure 4.18 Sand for sand filtration

Figure 4.19 Carbon filters and carbon filter pump

Figure 4.20 Micron filters

Figure 4.21 UV Disinfection: Sanitron™ Ultraviolet Purifier

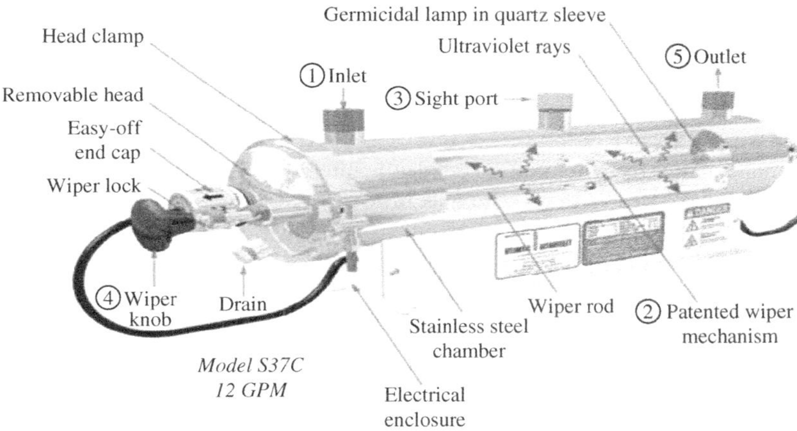

① The water enters the purifier and flows into the annular space between the quartz sleeve and the outside chamber wall.

② The wiper segments induce turbulence in the flowing liquid to assure uniform exposure of suspended microorganisms to the lethal ultraviolet rays.

③ Translucent sight port provides positive indication of germicidal lamp operation.

④ The wiper assembly facilitates periodic cleaning of the quartz sleeve without any disassembly or interruption of purifier operation.

⑤ Water leaving the purifier is instantly ready for use.

Figure 4.22 Process description for Sanitron™ Ultraviolet Purifier (Sanitron, 2005)

4.4.3 Wastewater treatment

4.4.3.1 Current situation

Currently, the wastewater from STRI is routed into five concrete septic tanks distributed throughout the station site. The septic tanks that carry wastewater from the laboratory building have a gravel drainfield, while the septic tank that carries wastewater from the dormitories does not have a drainfield. The original plan for the building was to recycle the wastewater through a constructed wetland system. The failure of one of the septic tanks, which leaked and polluted a nearby mangrove, helped to further motivate a better wastewater treatment on-site system.

4.4.3.2 Wastewater stabilization pond/constructed wetland plan

There is a current plan to build an integrated wastewater stabilization pond (WSP) and constructed wetland (CW) for the wastewater effluent at STRI. According to the United Nations Environment Program (UNEP), combining the use of constructed wetlands with waste stabilization ponds is an attractive wastewater treatment method for developing countries because (1) through the use of filter media such as sand or gravel, a CW can provide a significant reduction of suspended matter from maturation pond effluent, (2) a CW can buffer the pH value of the effluent from a WSP, (3) effluents from waste stabilization ponds most often need a post treatment polishing step and the use of wetlands offers an excellent cost-effective solution to this problem, and (4) the use of waste stabilization ponds and wetlands, in combination, offers a higher level of certainty than a WSP or CW individually as an effective pollution abatement and aquatic ecosystems management strategy (UNEP, 2003). See Figures 4.23a and 4.23b for a photograph and a diagram of the STRI site for the WSP/CW.

Local Panamanian engineering group, Caisa Inc., in conjunction with the Ministry of Health in Panama are in the design stages of the WSP/CW system. The integrated system is meant to treat the wastewater (both greywater and blackwater together). The area where the system will be constructed has 41 banana trees, 33 sugar canes, and 1 almond tree on the premises. The area that surrounds the project is a mangrove with trees that reach to 30 meters in height, and bushes of about 70 species that serve as the home for reptiles, birds, and mammals. Because part of the WSP/CW site was an old solid waste dump, it was decided that the WSP should be built above ground to avoid any contamination from hazardous chemicals which may exist on the dump site. The actual CW site, though adjacent to the solid waste dump, is not contaminated and was not used as a dump.

Figure 4.23a STRI site for WSP/CW

The system works as follows. The wastewater is pumped from various sources (bathrooms, kitchens, etc.) through sewer line pipes that lead to an above-ground WSP. At the end of the sewer line, which has pipes of 156 mm (6 inches) diameter, a pump for the wastewater will be installed to reach the level of the pond. A waterproof geomembrane will be placed in the bottom of the WSP to prevent water infiltration. The wastewater is allowed to aerate in the pond, and then the treated wastewater travels through pipes down to the CW on the ground. From the CW, the further-treated wastewater is finally transported to the sea. The tides of the Panamanian Caribbean Sea have one high tide and one low tide, or two high tides and two low tides in each period of 24 hours. The maximum fluctuation of the tide is approximately 0.4 m. The discharge of the treated wastewater to the sea will occur into the sand at a depth of 13 m; somewhere within 150 m of the sand, the sea will be encountered. Afterward, the sea meets a coral bed. The quality of the effluent that will be discharged in the permitted zone should indicate that the fauna of the area will not be affected (Deago, 2005). Figure 4.24a helps to illustrate the transport of the wastewater to the ocean; Figure 4.24b shows the shoreline at STRI.

The design flow of the system is estimated to be 24.2 m^3 per day (6,400 gallons per day). The pond will more than likely be built with earth. It will be designed with a retention period of 15 days, for a total volume of 364 m^3. It will

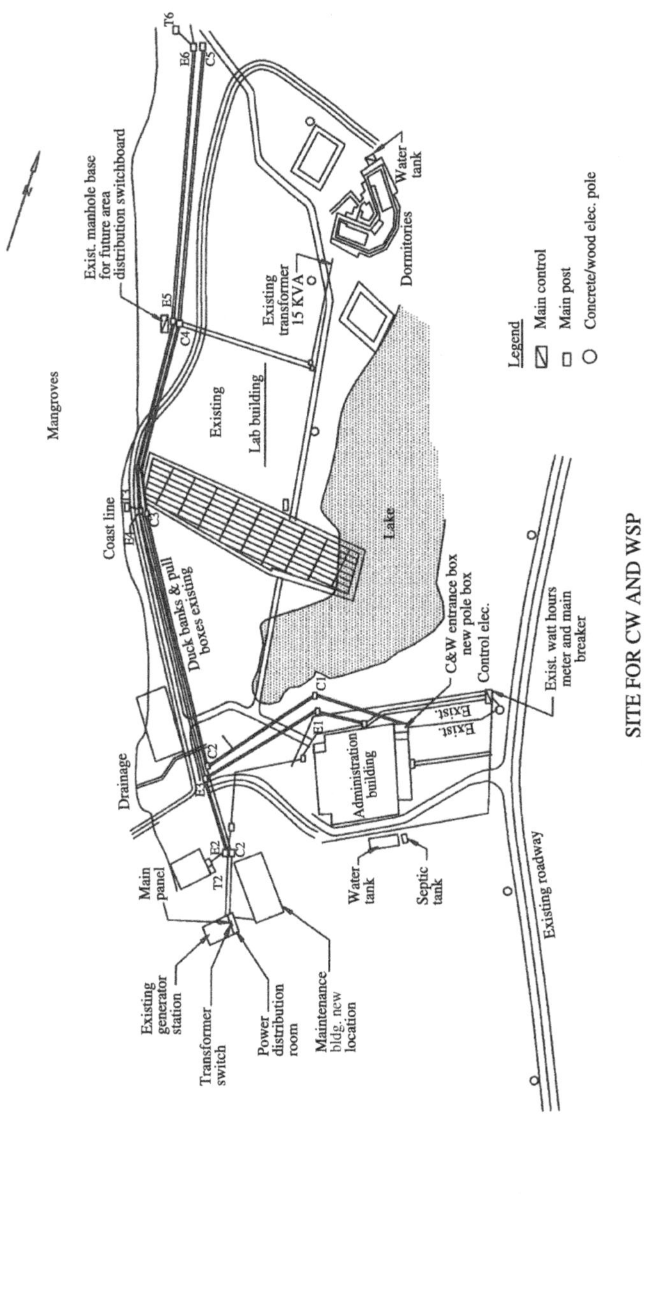

Figure 4.23b Site plan for STRI Bocas del Toro station; shows site for WSP and CW

be approximately 2 meters deep, to give a surface area of 182 m^2. The pond will thus be about 20 meters long and 10 meters wide. Additionally, it will have an area of 100 m^2 for the elimination of algae (Deago, 2005). See Figure 4.25 for a schematic diagram of the WSP.

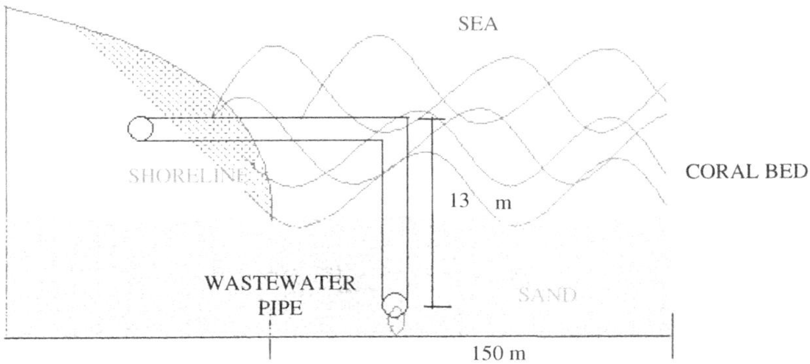

Figure 4.24a Transport of wastewater from WSP/CW system to sea

Figure 4.24b Shoreline for treated wastewater disposal

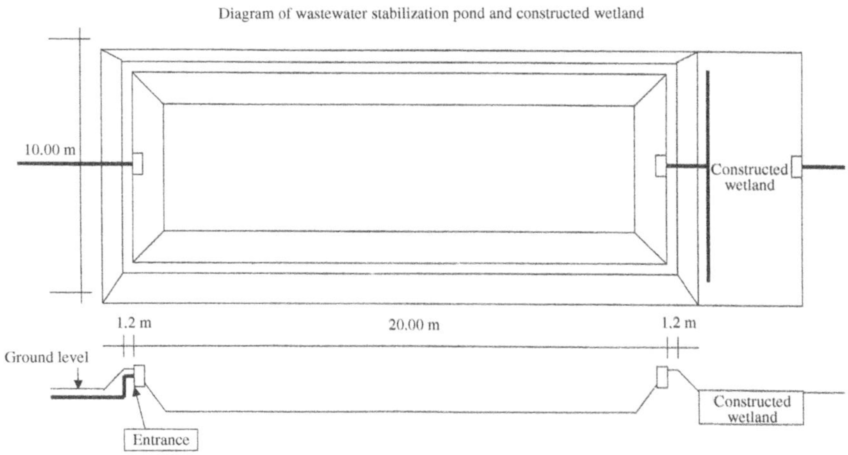

Figure 4.25 Plan for wastewater stabilization pond and constructed wetland at STRI (Deago, 2005)

The following stages of the WSP/CW construction have been planned. In the planning stage, the first task to be carried out is the leveling of the topography of the selected site. Soil will be taken from existing quarries on the island to produce about 100 m^3 of backfill for the site. The load capacity of the ground and the characterization of the wastewaters (hourly and daily flow variations, quality variations including chemical oxygen demand, biological oxygen demand, pH, oils and greases, nitrogen, phosphates, total dissolved solids, total suspended solids, total settleable solids, total solids, and coliform count) will be determined. The most adequate solution for the treatment of the wastewater will be designed. Table 4.1 shows the results of the wastewater quality tests that were completed at the entrance of the existing septic tanks from the main laboratory and from the machine workshop station.

The efficiency of the proposed system, based on the study of the existing literature, is established at 90% (this is a conservative number, and it could already be greater because of the subsequent treatment system to the lagoon). If a BOD of 300 mg/L is used as a base (that is higher than the observed) we will have a discharge of approximately 30 mg/L, which complies with the normal Panamanian discharge of wastewaters in water bodies. Equal consideration is made for the discharge of suspended solids. The existing mangrove swamp on-site is receiving the discharge from the existing septic tanks. The analysis from these discharges that unload into the mangrove swamp is presented in

Table 4.2. It is estimated from this table that inadequately treated wastewater is arriving to the mangrove swamp.

Table 4.1 Wastewater characterization at STRI, June 2005 (Deago, 2005)

	PARAMETER	WASTEWATER SAMPLE	
		Sample from the laboratory	Sample from the machine workshop station
1.	pH	6.73	7.03
2.	Total dissolved solids (mg/L)	455	509
3.	Suspended solids (mg/L)	47	41
4.	Total solids	502	550
5.	Sediment solids (mL/L)	0.14	0.42
6.	BOD_5 (mg/L)	102	181
7.	COD (mg O_2/L)	189	311
8.	Phosphates (mg PO_4/L)	11.3	13.4

Table 4.2 Characterization of wastewater going into mangrove Swamp at STRI, 2005 (Deago, 2005)

	PARAMETER	WASTEWATER SAMPLE IN THE MANGROVE SWAMP	
		Sample 1	Sample 2
1.	Organic Nitrogen (mg N/L)	0.32	0.11
2.	Total Nitrogen (mg/L)	0.32	0.11
3.	Fecal coliform	6×10^4	3.0×10^4
4.	Total coliform	8×10^4	5×10

During construction, a cottage will be constructed to keep the construction materials and equipment. Later the land will be redefined for the alignment of the drainage pipes and the correct placement of the WSP. With the help of a bulldozer, the excavation will be carried out for the placement of the pipes for the sewer system. The material from the excavation will be placed to one side of the trench to use as backfill. This material will be covered with a canvas to avoid erosion by the rain. The wastewater will be transported to the sea using pipes that have counterweights to guarantee that the pipes will stay submerged. A team of divers will verify that the pipes are positioned properly in the sea bottom.

The facilities will be maintained by the maintenance personnel of STRI. The main maintenance duties include ensuring that the roots of the trees and other types of vegetation do not penetrate the interior of the tubing through the connectors, keeping the wastewater flow free of obstacles, and eliminating any solid supernatant by means of manual removal. These solids will be placed in a pit that will be indicated by the WSP/CW design. In the same pit, the biological solids will be placed that are removed from the bottom of the lagoon, which will be done at least once per year. Additionally, the operation of the system will require removing the vegetation that grows in it. The lagoon will be designed to keep oxygen diffused throughout the entire column of water, so that anaerobic conditions do not develop and the area is kept free of unpleasant smells. The construction of the sewer system and WSP/CW has an investment cost estimated in 108,500 USD.

The breakdown of these costs (in USD) is as follows:

Sewer system: (300 m at a rate of 25 per m)	58,000
Treatment (Global)	38,000
Final Disposal (500 m at a rate of 10 per m)	12,500
Total	108,500

The CW will be made up of available native plants that will remove nitrates and phosphates from the wastewater. The vegetation that currently occupies the site for the CW are clumps of old mangroves and arbuscleous vegetation where one can find species of epiphytals and orchids (STRI, 2000). In most of the STRI building area, dominant species are of the family cyperaceae, with an abundance of *Elaecharis sp.* in stagnant waters of shallow depth and marshy land. There are rooted, submerged, and floating aquatic and amphibious plants (plants that can grow and develop in wet places and dry ones) such as *Jussea sp.*, *Canna sp.*, *Thalia sp.*, *Typha sp.*, and *Eleocharis sp. Typha sp.* are typical of wetlands and are characterized by colonization of large areas and fast reproduction (STRI, 2000). See Figure A.2 in the Appendix for common types of native plants used in a CW. The entire WSP/CW site is grassy and void of trees, with the exception of one tree which will not be removed.

In a conversation with Herbert Sedelmeier, engineer at STRI in Panama City, he stated that because of the maintenance problems in the Bocas del Toro municipal WSP, many are skeptical about the WSP and CW plan for the STRI station (Sedelmeier, 2005). The municipal WSP involves the same concepts as the WSP plan for the STRI building. The only difference is that the STRI pond will be built above ground and will have an added benefit of a CW. Because the local pond has not been maintained and currently causes odors, contamination problems, and is a generally inefficient wastewater treatment system, some wonder if the STRI pond will end up having the same problems. However, it is clear that the current septic tank system at STRI is not sustainable; a mangrove area on the site has already been contaminated from septic tank waste.

4.4.4 Water conservation methods

The plan for the STRI building originally included the use of composting toilets. In one of the early design development reports, it was stated that two central composting systems with full ventilation (vent and electric fan) would be installed. The composting toilet model recommended was a Clivus® company product showed in Figure 4.26.

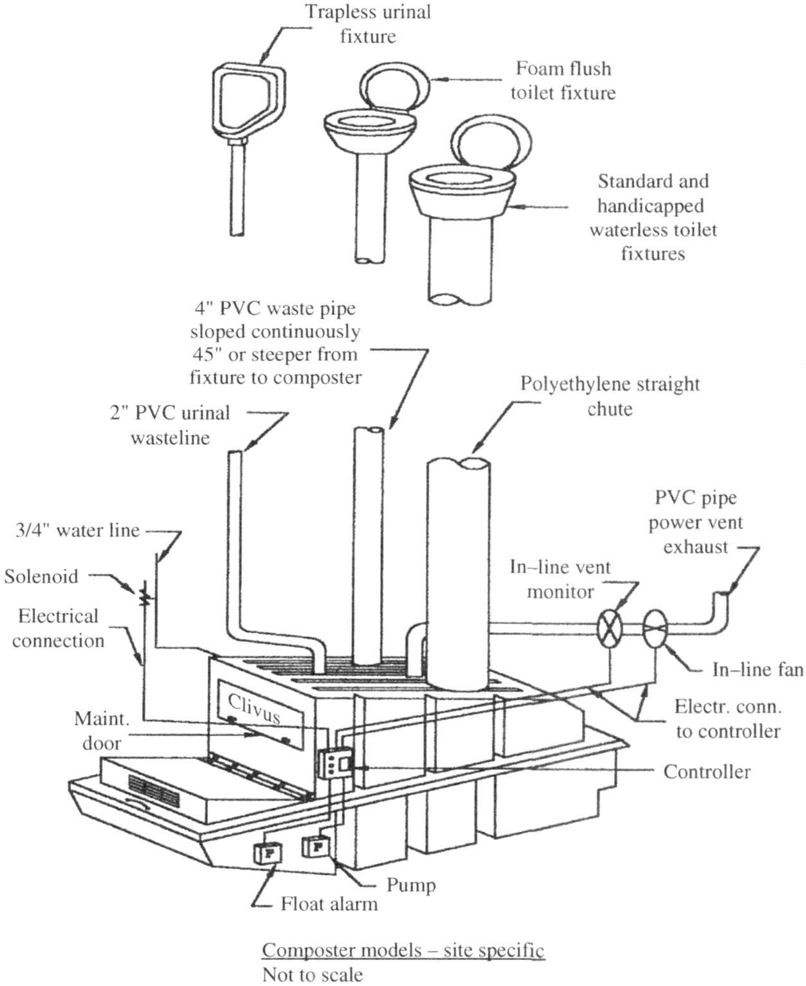

Figure 4.26 Clivus® composter models (Clivus, 2002)

However, it was decided that composting toilets would not be feasible because they would require too much maintenance. In addition, they were more expensive than traditional water flush toilets. Although composting toilets can be extremely economical, the ones considered for use in the STRI building were not, partly because of aesthetic embellishments. Initially, the STRI laboratory had installed low-flow water flush toilets that used 1.6 gallons (about 6 liters) of water per flush. Unfortunately, whenever there was a group of 20 or more people at the lab, there would be many complaints. According to Dennis Allen, engineer at STRI in Bocas, many people would misuse the toilets by placing paper towels for hand wiping in them (Allen, 2005). Keeping up the maintenance for the 1.6 gallon toilets soon became a significant problem, so STRI officials bought toilets that used 3.5 gallons (about 13.2 liters) per flush. These toilets, which used more than double the amount of water as before, were found to cause fewer complaints. It cost less than 500 USD to install four, 3.5 gallon toilets, while it cost approximately 800 USD to install four, 1.6 gallon toilets.

4.5 CONCLUSION

The water aspects of the STRI building aim to provide adequate wastewater and potable water systems in order to account for failures in the corresponding municipal systems. The municipal potable water for Bocas del Toro is pumped from Big Creek and undergoes physical and chemical water purification. Nevertheless, the municipal treatment system often produces unreliable water quality due to technical flaws and limited capacity for the current water demand. The municipal wastewater aeration pond, which is in very poor condition, services less than half of the town of Bocas del Toro. The remaining population either uses septic tanks that are improperly working or disposes of wastewater directly into the sea. The STRI laboratory building has a RWH system with a maximum storage capacity of 8,000 liters of water per day. The rainwater undergoes extensive treatment, which includes sand filtration, activated carbon filtration, micron filtration, and UV disinfection. Although there are currently five septic tanks that handle the wastewater from STRI, there is a plan to build a combined WSP and CW system. This system is meant to handle a wastewater volume of 24,200 liters per day. Initially, the STRI building was meant to use composting toilets to conserve water, but it was decided that the maintenance of the toilets would cause too many problems; water flush toilets are used instead. The RWH and WSP/CW system at STRI are two methods of incorporating sustainable water practices into a building. It remains to be seen how effective these systems are at maintaining an adequate quantity and quality of potable water, as well as properly treating and disposing of wastewater.

5
Water quality testing at Panama Site *The Smithsonian Tropical Research Institute Sustainable Building in Bocas del Toro*

5.1 INTRODUCTION: WATER QUALITY TESTING AT PANAMA SITE

5.1.1 Rainwater harvesting water quality experiment

In the rainwater harvesting water quality tests completed for Panama, the following parameters were tested: total suspended solids (TSS), turbidity, total and fecal coliform, sulfates, pH, color, nitrates, and nitrites. Some of these contaminants are more threatening to one's health than others, while others

ISBN: 9781843393238. Published by IWA Publishing, London, UK.

simply pose aesthetic effects. Water samples were taken from four locations: a sink faucet in the STRI building to test the municipal water, the three RWH storage tanks which were filled with rainwater, rain running off of the roof, and an overflow pipe. This chapter details the experimental set-up, procedures, and results of the RWH experiment.

5.1.2 Sampling sites and sample collection

The following water locations were tested in the water quality test experiments:

(1) *Municipal Water* from a sink in a bathroom at the STRI lab (Figure 5.1)

Trial 1: Sample Taken on Wednesday, November 2, 2005
Trial 2: Sample Taken on Friday, November 4, 2005

Figure 5.1 Municipal water from a sink in a bathroom at the STRI lab

(2) *Roof-water* that flows from a hole in the roof gutter (Figure 5.2)

Trial 1: Sample Taken on Wednesday, November 2, 2005
Trial 2: Sample Taken on Thursday, November 3, 2005
Trial 3: Sample Taken on Friday, November 4, 2005

Figure 5.2 Roof-water that flows from a hole in the roof gutter

(3) *Storage Rainwater* stored in three rainwater storage tanks (1000 gallons or about 3,785 liters each) (Figure 5.3)

Trial 1: Sample Taken on Thursday, November 3, 2005

Figure 5.3 Storage rainwater stored in three rainwater tanks

(4) Overflow Pipe Rainwater stored in one of the overflow pipes (see Rainwater Harvesting System Section) (Figure 5.4)

Trial 1: Sample Taken on Thursday, November 3, 2005

Figure 5.4 Overflow pipe rainwater

5.1.3 Materials

The following materials were used to complete the experimental methods for all tests: 1000 mL flasks, beakers, and bottles, 10 mL and 25 mL test tubes, 25 mL and 50 mL graduated cylinders, Decon Laboratories Contrad Cleaning Agent, latex gloves, goggles, clippers, tape, and a permanent marker. All samples were stored at 1–2°C if they could not be analyzed immediately. To complete the pH tests, a Baker-pHix wide ranging (pH 0.0–14) pH indicator stick kit and LaMotte octet comparator were used. The following materials were used to complete the TSS test: filter paper (5.5 cm VWR Brand), a balance (Daigger, DR-603D, Max 610/110 g, d = 0.01/0.001g), a vacuum Filter (100–115 V, Gast Model 0523-V4-G588DX, GE Motors, A-C Motor), an oven, and a desiccator. The following materials were used in the coliform tests: Coliscan Easygel® Easygel®, 20 mL bottles, pretreated petri dishes, 1 mL pipeter and pipet tips, ColiQuant EZ Manual, ColiQuant EZ Colony Color Guide, and an incubator.

For the colorimeter tests, the following were utilized: Sulfate SulfaVer® 4 Reagent Powder Pillows, Nitrate HR NitraVer® 5 Reagent, Nitrogen-Nitrite HR NitriVer® 2 Reagent, Hach DR/890 Colorimeter, DR/890 Datalogging Colorimeter Handbook, and Kimwipes®.

5.1.4 Methods

The following procedures were used for the water quality tests conducted in Panama. They are grouped into four major procedures: pH, total suspended solids (TSS), coliform, and parameters that required the use of a colorimeter. See Figures A.3 to A.8 in the Appendix for Experimental Set-ups.

5.1.4.1 pH

Two types of pH tests were used. One was using color-fixed indicator sticks and the other was using an octet comparator. A Baker-pHix wide ranging (pH 0.0–14) pH indicator stick kit was used to measure pH by simply placing a litmus paper stick in a water sample and matching the resulting color to set colors given by the kit. There are four color bands on each indicator stick, which will change color to match the appropriate pH (Mallinckrodt Baker, 2006). The LaMotte octet comparator has eight permanent color standards (pH 3–10), in which a test tube of the water sample, with indicator solution added, is compared in color to the eight standards (LaMotte, 2009). The following procedure was used:

(1) A test tube was filled to the 5.0 mL line with sample water.
(2) While holding the dropper bottle vertically, 10 drops of LaMotte WR Ind indicator solution (contains phenolphthalein, ethanol, and methanol) were added to sample water.
(3) The test tube was capped and mixed.
(4) The test tube was inserted in the Octet Comparator to match sample color to color standard. This was recorded as pH.

5.1.4.2 Total suspended solids (TSS)

Please refer to Table A.1 in the Appendix for the procedures used for the TSS test.

5.1.4.3 Total and fecal coliform test

Explanation of Coliscan Easygel EZ method. Coliforms produce certain enzymes in order to ferment lactose. These enzymes can be identified and used to verify the presence of coliforms and *Escherichia coli* (E. coli). General coliforms produce the enzyme *galactosidase* from lactose fermentation; E. coli, *or fecal coliforms,* produce the enzyme *glucuronidase* in addition to galactosidase.

Coliscan Easygel incorporates two special chromogenic substrates, which are acted upon by the enzymes galactosidase and glucuronidase, to produce pigments of different colors. General coliforms produce the enzyme galactosidase, with colonies that grow in a pink color. E. coli produces galactosidase and glucuronidase, with colonies that grow in a blue/purple color. *Total coliforms* can be defined as the sum of the E. coli (known as fecal coliforms for the Coliscan Easygel test) and general coliforms (LaMotte, 2009).

The following describes important definitions in the coliform tests. Colony-forming units are represented by the abbreviation CFU. A CFU can be seen without magnification, and may contain more than 10,000 individual bacterial cells (all clones of the original bacterial cell). E. coli are those bacteria which grow as dark blue/purple colonies on the Coliscan Easygel plate as a result of the production of both glucuronidase and galactosidase enzymes. These bacteria are of fecal origin and are known as *fecal coliforms*. Total Coliforms are those bacteria which make up the sum of the E. coli (blue/purple colonies) + general coliforms. The general coliforms will appear as pink/magenta colonies because they produce galactosidase, but NOT glucuronidase. A color guide example for identifying colonies is provided in Figure A.9 of the Appendix. Species of the genera Citrobacter, Enterobacter, Escherichia, and Klebsiella are the main groups of coliform bacteria. Non-Coliform are bacteria that form colonies that are not blue/purple or pink/magenta on Coliscan Easygel medium and are not considered to be of fecal origin. They live naturally in soil and water.

Coliform procedure. Please refer to Table A.2 in the Appendix for the procedures used for the coliform tests.

5.1.5 Hach methods (Hach DR/890 Colorimeter)

This section covers the Hach DR/890 Colorimeter methods for true and apparent color, nitrates, nitrites, sulfates and turbidity. Please note that all sample cells were wiped with Kimwipes when necessary to remove any fingerprints, smudges, or liquid on test tubes (samples cells) and all test tubes were thoroughly cleaned after each use.

5.1.5.1 Precision and detection limit

For each colorimeter parameter tested, the precision level and detection limit for the procedure are given. Every measurement has some degree of uncertainty. The quality of the entire chemical method determines the precision. Uncertainty in chemical measurements may be due to systematic errors and/or random errors. Hach provides an estimate of each procedure's precision. The procedures use a "replicate analysis" estimate, based on real data (Hach, 2005). In replicate analysis, a Hach chemist prepares a specific concentration of the analyte in a deionized water matrix. The standard is then analyzed seven individual times with the two reagent lots used in the calibration (14 total samples). A standard deviation of the two sets of seven values is calculated. The larger value is reported in the method. The reported value provides an estimate of the "scatter" of results at a particular point in the calibration curve.

Ranges for chemical measurements have limits. The lower limit is important because it determines whether a measurement is different from zero. The method detection limit (MDL) is the lowest concentration that is different from zero with a 99% level of confidence. A measurement below this MDL may be useful, but there is a greater chance that it is actually zero. The MDL is not fixed; it varies for each reagent lot, instrument, analyst, sample type, etc. Therefore, a published MDL may be a useful guide, but is only accurate for a specific set of circumstances. Hach provides a value called the Estimated Detection Limit (EDL) for all programs. It is the calculated lowest average concentration in a deionized water matrix that is different from zero with a 99% level of confidence. Specifically, it is the upper 99% confidence limit for zero concentration based on the calibration data used to prepare the pre-programmed calibration curve.

5.1.5.2 Color, true and apparent

Method. APHA Platinum-Cobalt Standard (0 to 500 units). Please refer to Table A.3 in the Appendix for the procedures used for the true and apparent color tests.

Precision and detection limit. In a single laboratory, using a standard solution of 250 Pt-Co color units and two representative lots of reagent with the instrument, a single operator obtained a standard deviation of ± 10 Pt-Co color units. The estimated detection limit (EDL) for program 19 is 25 Pt-Co color units.

Summary of method. Color may be expressed as "apparent" or "true" color. The apparent color includes color from dissolved materials plus that from suspended matter. By filtering or centrifuging out the suspended materials, the true color can be determined. The procedure describes true color analysis. If apparent color is desired, it can be determined by measuring an unfiltered water sample. The stored program is used for both forms of color.

5.1.5.3 Nitrate

Method. Cadmium Reduction Method (0 to 30.0 mg/L (NO_3^--*N*). Please refer to Table A.4 in the Appendix for the procedures used for the nitrate tests.

Precision and detection limit. In a single laboratory using standard solutions of 25.0 mg/L nitrate nitrogen (NO_3^--*N*) and two representative lots of reagent with the instrument, a single operator obtained a standard deviation of ± 1.7 mg/L nitrate nitrogen for program # 51. The EDL for program 51 is 0.8 mg/L NO_3^--*N*.

Summary of method. Cadmium metal reduces nitrates present in the sample to nitrite. The nitrite ion reacts in an acidic medium with sulfanilic acid to form an intermediate diazonium salt which couples to gentisic acid to form an amber-colored product.

5.1.5.4 Nitrite

Method. Ferrous Sulfate Method (0 to 150 mg/L (NO_2^-)). Please refer to Table A.5 in the Appendix for the procedures used for the nitrite tests.

Precision and detection limit. In a single laboratory using a standard solution of 123 mg/L nitrite and two representative lots of reagents with the instrument, a single operator obtained a standard deviation of +1 mg/L nitrite. The EDL for program 59 is 2 mg/L (NO_2^-).

Summary of method. The method uses ferrous sulfate in an acidic medium to reduce nitrite to nitrous oxide. Ferrous ions combine with the nitrous oxide to form a greenish-brown complex in direct proportion to the nitrite present.

5.1.5.5 Sulfate

Method. SulfaVer 4 Method (0 to 70 mg/L). Please refer to Table A.6 in the Appendix for the procedures used for the sulfate tests.

Precision and detection limit. In a single laboratory, using a standard solution of 50 mg/L sulfate and two representative lots of powder pillows with the instrument, a single operator obtained a standard deviation of ± 0.5 mg/L sulfate. The EDL for program 91 is 4.9 mg/L SO_4.

5.1.5.6 Turbidity

Method. Absorptometric Method (0 to 1000 FAU). Please refer to Table A.7 in the Appendix for the procedures used for the turbidity tests.

Description of turbidity test. The turbidity test used measures an optical property of the sample which results from scattering and absorption of light by particles in the sample. The amount of turbidity measured depends on variables such as the size, shape, color, and refractive properties of the particles. This procedure was calibrated using formazin turbidity standards and the readings are in terms of Formazin Attenuation Units (FAU). One FAU is equivalent to one Nephelometric Turbidity Unit (NTU) of Formazin. However, the optical method of measurement for FAU is very different than the NTU method (1 NTU = 1 FTU = 1 FAU when traced to formazin primary standards.)

Precision and detection limit. In a single laboratory, using a turbidity standard solution of 200 FAU with the instrument, a single operator obtained a standard deviation of ±2 FAU. The EDL for program 95 is 21 FAU.

5.2 RAINWATER HARVESTING EXPERIMENT RESULTS

5.2.1 Coliform results

Table 5.1 presents the fecal and total coliform results for the four water locations tested. See Table A.8 in the Appendix for a complete spreadsheet of the calculations for coliforms. See Table A.9 in the Appendix for a statistical analysis of the Coliform data where appropriate. The sample of municipal water for Bocas del Toro had no detectable fecal and total coliforms. However, because only one trial was completed, it is uncertain whether this represents the general quality of the municipal water in regards to coliforms.

The sample of rain from the roof on Day 1 had 0 fecal coliforms and an average of 190 CFU/100 mL of total coliforms; on Day 2 the sample of rain from the roof had an average of 520 CFU/100 mL of fecal coliforms and an average of 620 CFU/100 mL of total coliforms. This indicates that the presence of coliforms in rainwater from a roof will vary from day to day, depending on the conditions of that day. For example, bird droppings (which contribute to fecal contamination) will vary, pollutants on the roof will vary, and there will be more pollutants running off of the roof after a dry period. As explained in Chapter 1 in the Rainwater Harvesting section, contaminants from roof-water will be lessened after the first flush. Thus, it is better to collect rainwater from the roof after the roof has been cleaned with rain at least once. The roof will become more contaminated as the time without rain increases (dry period).

The roof-water sample on Day 1 was taken after the first flush (after the roof had been cleaned with rainwater at least one time during the same downpour); the roof-water sample on Day 2 was taken approximately ten minutes after it started raining and it was not clear if the first flush had taken place. This fact could also contribute to a greater amount of coliforms on Day 2.

Table 5.1 Total and fecal coliform results at STRI in Bocas del Toro, Panama (October 2005) amount of sample used (mL): 5 CFU: Colony forming units

Source	Trial #/storage tank #	Fecal coliform (CFU/100 mL)	Total coliform (CFU/100 mL)
Municipal Water (from Sink at STRI lab)	**Trial 1**		
	Petri Dish 1	0	0
	Petri Dish 2	0	0
Rain from Roof	**Trial 1**		
	Petri Dish 1	0	140
	Petri Dish 2	0	240
	Trial 2		
	Petri Dish 1	660	860
	Petri Dish 2	380	380
Rain Stored in Rainwater Storage Tanks	**Storage Tank 1**		
	Petri Dish 1	20	440
	Petri Dish 2	120	440
	Storage Tank 2		
	Petri Dish 1	260	460
	Petri Dish 2	80	500
	Storage Tank 3		
	Petri Dish 1	280	320
	Petri Dish 2	80	280
Rain from Overflow Pipe	**Trial 1**		
	Petri Dish 1	440	640
	Petri Dish 2	0	960

For rainwater in the three storage tanks, the coliform values effectively represent the average coliform values from the roof-water over time and space. Although there can be a short-term increase in bacterial numbers in the tank as the microorganisms take up residence, typical die-off behavior for microorganisms in water follows the pattern of a short period where numbers remain constant followed by exponential decline in numbers (DTU, 2005b). The rain from the roof goes to the storage tanks by way of pipes; it does not matter whether it is the first flush or not. This means that the water stored in the storage tanks should be representative of roughly an average of what is coming in from the roof on various days. The results support this. The average fecal coliform amount was 140 CFU/100 mL, and the average total coliform amount was 407 CFU/100 mL.

For the sample of rainwater in the overflow pipe, the average coliform values were higher than the stored rainwater, rainwater from the roof, and the municipal water. The average fecal coliform value was 440 CFU/100 mL, and the average total coliform value was 800 CFU/100 mL. The rainwater trapped in this pipe also varies from rainfall event to rainfall event. It shows a mixture of various rainfall events (less distributed than the three storage tanks).

The WHO and EC guideline for coliforms is 0 CFU/100 mL. The USEPA guideline for coliforms is less than 1 CFU/100mL. In comparison to the WHO, EC, and USEPA standards for coliform, the rainwater collected from the STRI building is not suitable for drinking unless it is treated. Table 5.1 shows the fecal coliform and total coliform count for water from (1) the municipality, (2) the rainwater running off of the roof, (3) the rainwater stored in the three storage tanks, and (4) the rainwater from an overflow pipe.

5.2.2 Results for all other parameters

Table 5.2 shows results for the four water locations tested for the following water quality parameters: pH, TSS, turbidity, sulfates, nitrates, nitrites, apparent color, and true color. See Figure A.8 in the Appendix for a statistical analysis of the TSS data. Sulfate, nitrate, and pH values were in the acceptable limits of WHO, USEPA, and EC drinking water standards. However, some of the nitrite, turbidity, and color values were not within WHO, USEPA, and EC standards. There are no WHO, USEPA, or EC standards for TSS. The fact that the nitrite, turbidity, and color values exceeded the limits set by the WHO, USEPA, and EC, further indicates that the rainwater should be treated before it is used as a potable water source.

Table 5.2 Results of all other parameters tested at STRI in Bocas del Toro, Panama (October 2005) Note: – indicates not tested; ND: Non-detectable; EDL: estimated detection limit

Parameter	Source			
	Roof-water	Storage Rainwater	Municipal Water	Overflow Water
ph (Octet Comparison)				
Trial 1/Storage Tank 1	6	6	8	6
Trial 2/Storage Tank 2	6	6	8	–
Trial 3/Storage Tank 3	6	6	–	–
ph (Litmus test)				
Trial 1/Storage Tank 1	8	7.5	7.5	7.5
Trial 2/Storage Tank 2	7.5	7.5	7.5	–
Trial 3/Storage Tank 3	7.5	7.5	–	–
Total Suspended Solids (mg/L)				
Trial 1/Storage Tank 1				
(Trial a)	40	0	40	70
(Trial b)	20	10	10	60
(Trial c)	40	0	30	110
Trial 2/Storage Tank 2				
(Trial a)	30	20	–	–
(Trial b)	50	20	–	–
(Trial c)	30	50	–	–
Trial 3/Storage Tank 3				
(Trial a)	–	20	–	–
(Trial b)	–	10	–	–
(Trial c)	–	30	–	N/A
Turbidity (FAU) [EDL: 21 mg/L]				
Trial 1/Storage Tank 1	0 (ND)	3 (ND)	2 (ND)	0 (ND)
Trial 2/Storage Tank 2	3 (ND)	2 (ND)	0 (ND)	–
Trial 3/Storage Tank 3	1 (ND)	0 (ND)	–	–
Sulfates (mg/L) [EDL: 4.9 mg/L]				
Trial 1/Storage Tank 1	0 (ND)	0 (ND)	51	0 (ND)
Trial 2/Storage Tank 2	0 (ND)	1 (ND)	58	–
Trial 3/Storage Tank 3	0 (ND)	0 (ND)	–	–
Nitrate**s (mg/L) [EDL: 0.8 mg/L]**				
Trial 1/Storage Tank 1	0.6 (ND)	1.1	0.7	1.1
Trial 2/Storage Tank 2	1.1	0.9	1.3	–
Trial 3/Storage Tank 3	0.8	0.7 (ND)	–	–
Nitrites (mg/L) [EDL: 2 mg/L]				
Trial 1/Storage Tank 1	0 (ND)	1 (ND)	1 (ND)	1 (ND)
Trial 2/Storage Tank 2	1 (ND)	1 (ND)	1 (ND)	–
Trial 3/Storage Tank 3	3	2	–	–

(*continued*)

Table 5.2 *Continued*

Parameter	Source			
	Roof-water	Storage Rainwater	Municipal Water	Overflow Water
True Color (Pt-Co Units) [EDL: 25 mg/L]				
Trial 1/Storage Tank 1	0 (ND)	0 (ND)	21 (ND)	0 (ND)
Trial 2/Storage Tank 2	0 (ND)	0 (ND)	–	–
Trial 3/Storage Tank 3	–	1 (ND)	–	–
Apparent Color (Pt-Co Units) [EDL:25 mg/L]				
Trial 1/Storage Tank 1	6 (ND)	0 (ND)	71	10 (ND)
Trial 2/Storage Tank 2	0 (ND)	1 (ND)	–	–
Trial 3/Storage Tank 3	–	4 (ND)	–	–

5.3 SUMMARY

This chapter provided information regarding the water quality experiments completed at the STRI laboratory building in Bocas del Toro, Panama. The following water sources were tested for various chemical and biological contaminants: the rainwater from three installed storage tanks, the rainwater running off of the roof during three different downpours, the rainwater stored in one of the overflow pipes, and the municipal potable water taken from a sink in the building. The parameters tested include TSS, turbidity, pH, sulfates, color, nitrates, nitrites, and total and fecal coliform. The most important result from this study is that both the rainwater collected directly from roof runoff and the rainwater collected from the three storage tanks were contaminated with coliforms. Thus, one should not assume that rainwater collected from a catchment surface is a clean source. The data also shows that the quality of the rainwater varies depending on the current conditions (of the roof, wildlife activities, climate, etc.) since the quality of the rainwater from the roof changed in each trial. It is therefore critical to take appropriate measures to prevent, reduce, and treat contaminants in rainwater.

6
Kenya Site and Water Management Practices
The village homes near the Mpala Research Centre Laikipia, Kenya

6.1 INTRODUCTION: KENYA SITE AND WATER MANAGEMENT PRACTICES

This chapter presents information regarding the water aspects of the buildings at the Mpala Research Centre (MRC) and its surrounding communities in Laikipia, Kenya. In particular, this chapter provides a description of Kenya climate and water resources, the MRC study area used for field work and water quality tests, and water management at MRC and its surrounding communities. Data was collected by interviewing officials and community

ISBN: 9781843393238. Published by IWA Publishing, London, UK.

members, as well as by performing water quality experiments on rainwater and local river water. The rainwater samples were tested for total coliform, fecal coliform, sulfates, nitrates, nitrites, pH, apparent color, and turbidity in order to determine (1) the suitability of RWH for drinking purposes and (2) what kind of treatment (if any) is needed. Local river resources were also tested for the same parameters. Chapter 8, Water Quality Testing at Kenya Site, gives a description of the types of tests that were conducted and presents the findings.

6.2 DESCRIPTION OF KENYAN CLIMATE AND WATER RESOURCES

The Kenya case study was completed in the area surrounding the village of the MRC, which is located in the Central Province, in the Laikipia District of Kenya, northwest of Mount Kenya, 50 kilometers north of the Equator, and 50 kilometers from the town of Nanyuki. Kenya is located in East Africa with bordering countries Ethiopia, Somalia, Sudan, Tanzania, and Uganda, as well as the Indian Ocean bordering its southeastern side. Figures 6.1 and 6.2 illustrate a map of this region.

Most of Kenya (over 75%) is classified as semi-arid; only around 20% of its land is viable for agriculture. Kenya has a total area of 580,367 sq. km, an area covered by water of 11,227 sq. km, and a land area of 569,140 sq. km (CIA, 2009). The climate is generally warm and humid at the coast, cool and humid in the central highlands, and hot and dry in the north and east. Much of Kenya experiences heavy rainfall from March through May and to a lesser extent, from October through December. The average rainfall is 567 mm per year (IEA and SID, 2001). Figure 6.3 shows a map of the repeatedly drought affected areas of Africa, with Kenya being shaded as one.

With a country population of roughly 30 million, Kenya requires large supplies of water for its agricultural (65% of water supply), industrial (13% of water supply), and domestic (18% of water supply) activities (IEA and SID, 2001). Most of the Kenyan population, about 24 million people, gain their livelihood from land-based natural resources as farmers, herders, hunter-gatherers, or quasi-industrialists (IEA and SID, 2001). Unfortunately, drought, settlement in marginal areas, threats from pollution, siltation, reclamation, pesticides, weed attack, and human population activities has put immense pressure on water supply.

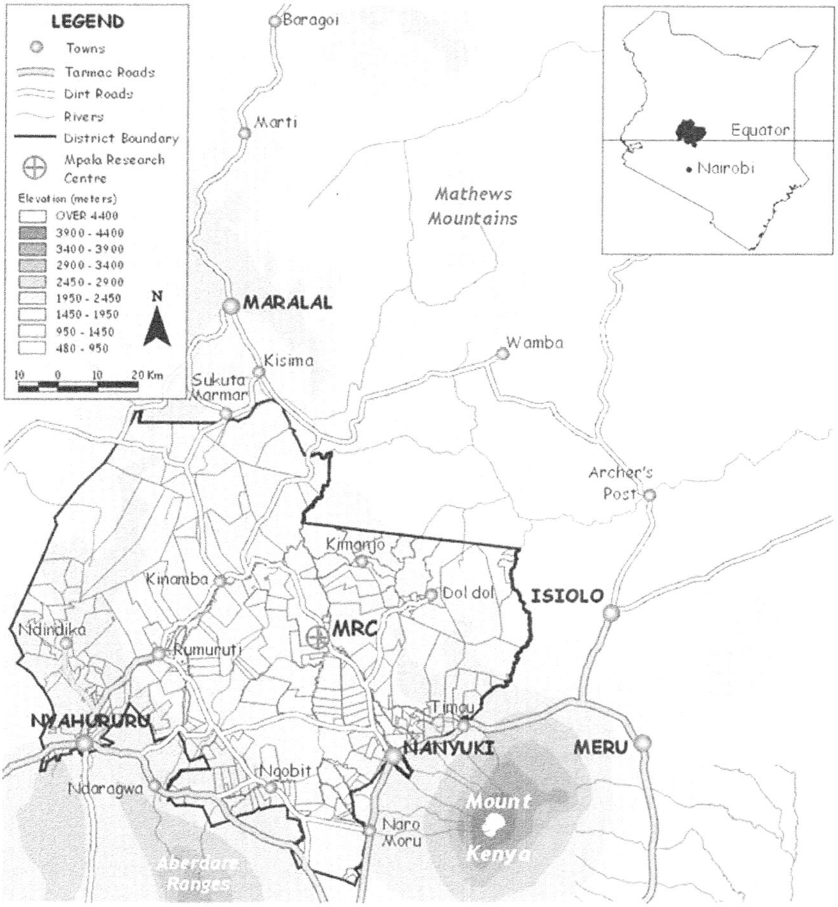

Figure 6.1 Map of Mpala Research Centre in Kenya (Mpala Wildlife Foundation, 2005)

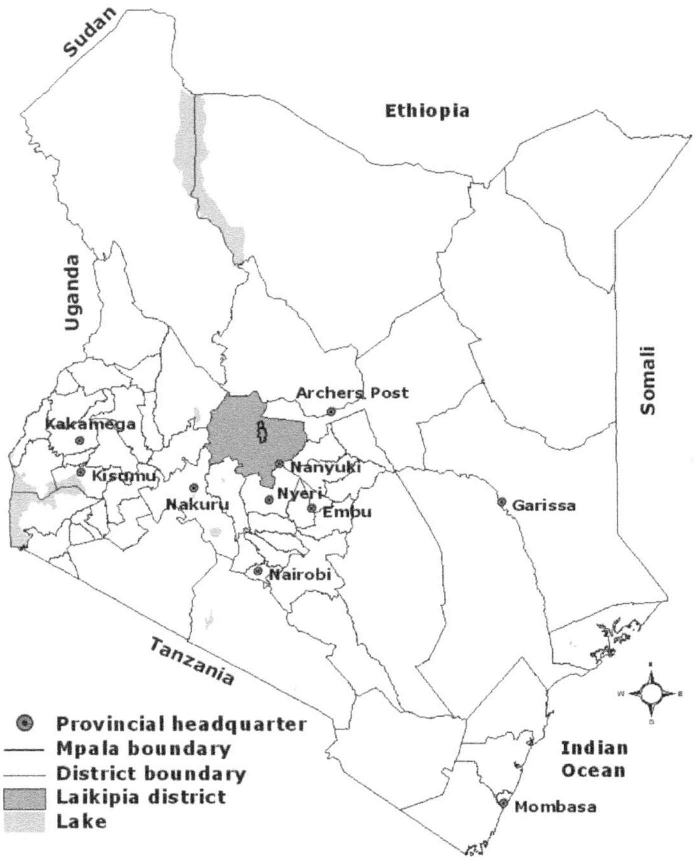

Figure 6.2 Map of Mpala Research Centre in Kenya (Mpala Wildlife Foundation, 2009)

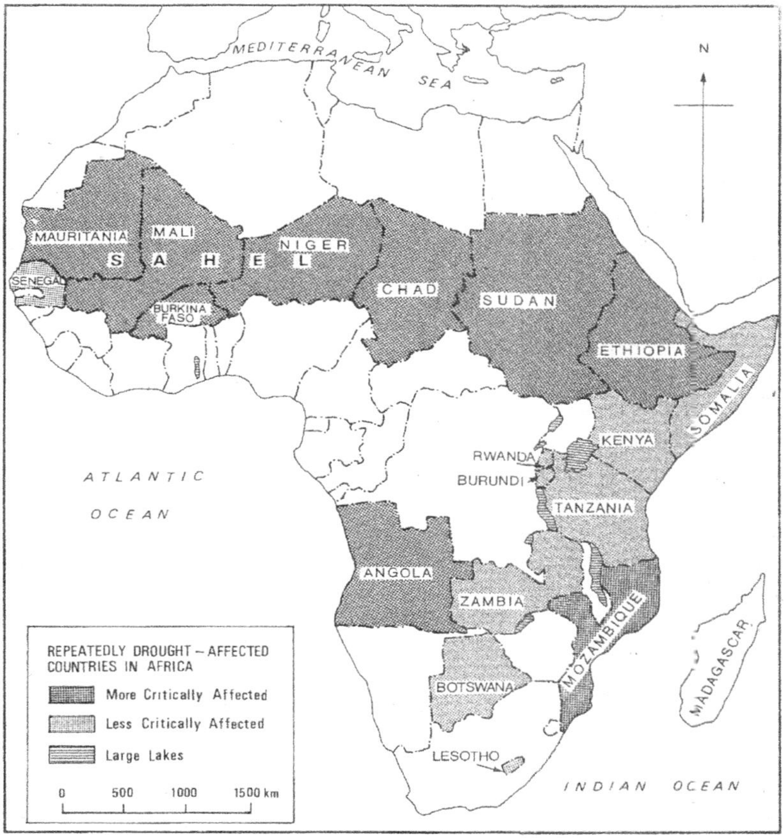

Figure 6.3 Drought-affected countries in Africa (Gupta, 1998) From: Ecology and development in the third world, A. Gupta, Copyright 1998 Taylor & Francis. Reproduced by permission of Taylor & Francis Books UK

6.3 DESCRIPTION OF STUDY AREA

The study area for this project is the MRC and surrounding villages in the Laikipia District of Kenya. The MRC is a scientific research center whose goals include helping to preserve and protect the Laikipia District's land, people, wildlife, and natural resources. Much of this area consists of semi-nomadic tribes, including communities such as the Maasai and Samburu. Livelihoods are partly maintained in the area by wheat production, livestock ranching, conservation research and activities, and eco-tourism. The MRC is bounded by two rivers, the Ewaso Nyiro (also Ng'iro) on the east side and the Ewaso Narok on the north side. The Ewaso Nyiro basin belongs to the hydrology for environmental life and policy (HELP) program. HELP is a joint initiative of UNESCO and the World Meteorological Organization (WMO); its aim is to form a stakeholder-driven approach to integrated catchment management (hydrological research and water-related observations) (Shuttleworth, 2007).

The study area suffers from destruction to natural vegetation, overgrazing, drought, and desertification. In general, Kenya has two major rainy seasons, the March to May "long rains" and the October to December "short rains" (Camberlin and Wairoto, 1997). Near MRC, however, the rainfall is "weakly trimodal" with peaks from April to May, July to August, and October to November, with a distinct dry season from January to February (Mpala Wildlife Foundation, 2003). The MRC property spans an altitudinal and climatic gradient that is higher, wetter, and cooler in the southwest (altitude of 1,850 m, average annual rainfall of 530 mm), and lower, drier, and hotter in the northeast (altitude of 1,550 m, average annual rainfall of 400 mm) (Mpala Wildlife Foundation, 2003). Fifteen dams have been constructed near the MRC, mainly for the purposes of livestock drinking and to help collect rainwater to be pumped into storage tanks. Figures 6.4 to 6.5 show rainfall patterns during the long and short rainy seasons for the year 2003.

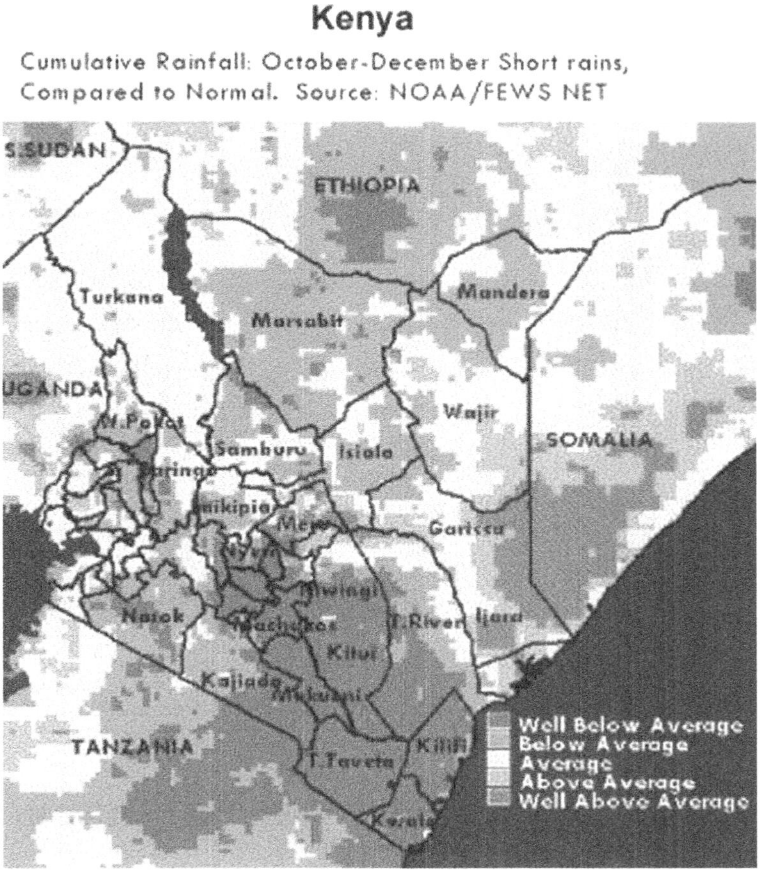

Figure 6.4 Cumulative rainfall for short rainy season 2003 compared to average (ReliefWeb, 2003a)

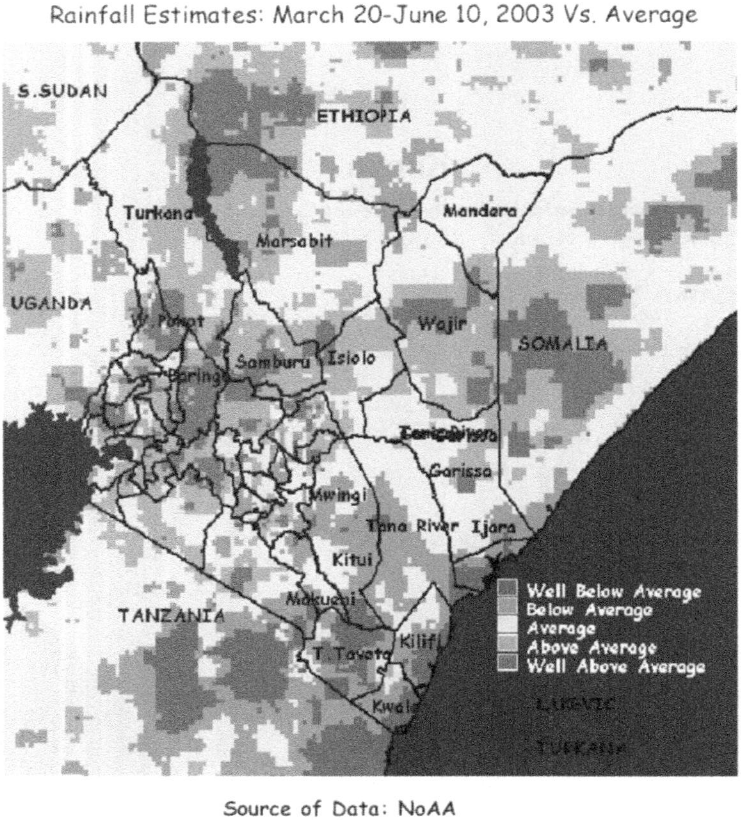

Figure 6.5 Cumulative rainfall for long rainy season 2003 compared to average (ReliefWeb, 2003b)

6.4 WATER MANAGEMENT IN LAIKIPIA NEAR THE MPALA RESEARCH CENTRE

6.4.1 Mpala Research Centre (MRC)

6.4.1.1 Water at MRC

A survey report on water supply and consumption was completed at the MRC in March of 2005. In this report, it is estimated that there are on average 510 people residing on the total Mpala property (190 at MRC and 320 at the Mpala

Conservancy), and each consumes about 89 liters of river water each day for general cleaning purposes. Each person is estimated to consume about 8 liters a day of *rainwater*, which is used for drinking, brushing teeth, and cooking. Water from the river is pumped uphill to a 23,000 L tank, where it flows downwards by gravity to various faucets. Since water for cattle is important in this area, cattle consumption of water was estimated at 40 liters per day for a total of 3,500 cattle. Tables 6.1 and 6.2 present the river water and rainwater consumption for the MRC and the Mpala Conservancy (Odhiambo, Petters and Wreford, 2005).

Table 6.1 River water utilization at the Mpala Properties (Odhiambo, Petters and Wreford, 2005)

(i) Human

	Centre	Conservancy	Total
Total consumption per day (L)	16,910	28,480	45,390
Total consumption per month (L)	524,210	882,880	1,407,090
Total consumption per year (L)	6,172,150	10,395,200	**16,567,350**

(i) Livestock

	Total
Consumption per day (L)	140,000
Consumption per month (L)	4,340,000
Consumption during drought (100 days) (L)	**43,400,000**

Note: The total river water requirement for human and livestock is 59,967,350 liters in a year (excluding wildlife).

Table 6.2 Rainwater consumption requirement rates at Mpala (Odhiambo, Petters and Wreford, 2005)

	Centre	Conservancy	Total
Consumption per day (L)	1,520	2,560	4,080
Consumption per month (L)	47,120	79,360	126,480
Consumption per year (L)	554,800	934,400	**1,489,200**

The MRC collects rainwater from thirteen buildings. The roofs of all buildings used for RWH are made out of corrugated and galvanized iron (GI) sheets. The harvested rainwater is used for potable purposes once it is boiled and ceramic filtered. The rainwater tanks are cleaned out once a year, using disinfectants, before the first rainy season begins. Table 6.3 shows the number and capacity of existing RWH storage tanks at the MRC and Mpala Conservancy.

Table 6.3 Existing water storage tanks at (a) MRC and (b) Mpala conservancy including their grand total (Odhiambo, Petters and Wreford, 2005)

Location	Existing tanks	Approx. Capacity (L)	Total Capacity (L)
(a) Centre			
Library	2	13,000	26,000
NSF Lab	1	13,000	13,000
McCormack Lab	3	13,000	39,000
Admin Block	2	6,000	12,000
Top Kitchen	1	2,000	2,000
Kitchen	2	7,450	14,900
Mess	1	2,000	2,000
Store 15	1	14,000	14,000
Jenga House	1	6,900	6,900
Director's House	1	7,000	7,000
Admin's House	1	6,900	6,900
GIS Analyst's House	1	13,000	13,000
Res. Scientist's House	1	13,000	13,000
Subtotal	**18**		**169,700**
(b) Conservancy			
Ranch House	2	9,500	19,000
Subtotal	2		**19,000**
Grand Total	**19**		**188,700**

Table 6.4 shows the amount of rainfall that can be harvested at two efficiencies (two run-off coefficients) of 0.7 and 0.9 (see Chapter 8 for a discussion of runoff coefficient). Mean annual rainfall is derived from climate stations between January 2000 and December 2004. During this period mean annual rainfall was 623.2 mm at the MRC and 431.8 mm at the Mpala Conservancy. During the driest year 2000, annual rainfall varied from 343.9 mm and 177 mm at the MRC and Mpala Conservancy, respectively.

The rainwater harvesting potential calculated in the Mpala survey was only for existing structures. There are dormitories and homes for long-term research staff on-site that contain thatch roofs for aesthetic purposes. If these were transformed to GI roofs, more rainwater could be harvested at MRC.

Table 6.4 Rainwater harvesting potential at Mpala (Odhiambo, Petters and Wreford, 2005)

Building	Roof Area Catchment (sq. m)	Mean annual rainfall		Minimum total rainfall (2000)	
		90% efficiency	70% efficiency	90% efficiency	70% efficiency
(a) Centre					
Admin Block	112.0	62,818.6	48,858.9	34,665.1	26,961.8
McCommack's Lab	175.0	98,154.0	76,342.0	54,164.3	42,127.8
NSF Lab	175.0	98,154.0	76,342.0	54,164.3	42,127.8
Library	199.0	111,615.1	86,811.8	61,592.5	47,905.3
Mess	286.0	160,411.7	124,764.6	88,519.9	68,848.8
Kitchen	66.0	37,018.1	28,791.8	20,427.7	15,888.2
Director's House	130.0	72,914.4	56,711.2	40,236.3	31,294.9
GIS House	155.0	86,936.4	67,617.2	47,974.1	37,313.2
Chris House	155.0	86,936.4	67,617.2	47,974.1	37,313.2
Admin House	90.0	50,479.2	39,261.6	27,855.9	21,665.7
Jenga House	175.0	98,154.0	76,342.0	54,164.3	42,127.8
Work Shop	120.0	67,305.6	52,348.8	37,141.2	28,887.6
Store 15	65.0	36,457.2	28,355.6	20,118.2	15,647.5
Subtotal rainwater (L)		**1,067,355**	**830,165**	**588,998**	**458,109**
(b) Conservancy					
HQ	122.0	47,411.6	36,875.7	19,435.7	15,115.7
Subtotal rainwater (L)		**47,412**	**36,876**	**19,436**	**15,116**
Grand total rainwater potential (L)		**1,114,766**	**867,040**	**608,433**	**473,225**

An independent study on water supplies for MRC was conducted by a private engineering firm in Kenya in November of 2005. The firm made the following recommendations among others: the intake from the Ewaso Nyiro River to the pumps should be improved, the performance of the existing pumps should be ascertained so that it can be decided whether it is viable to change them, a water clarification plant should be constructed near the pump house designed so that filtration and sterilization can be added later if found necessary, and the advantages of additional and improved rainwater harvesting should be investigated. Figures 6.6 to 6.8 were presented as a design for a river water treatment plant system. The system involves clarification through sedimentation, filtration through rapid sand filtration, and disinfection through chlorination. The river water that enters the system would flow over a perforated metal basket containing alum. The firm recommended that the best type of sedimentation tank system would involve two separate tanks with a capacity of 20,000 liters each (depth of 1.8 m, diameter of 4 m). Rapid sand filtration would involve a filter with diameter of 2 ft (600 m).

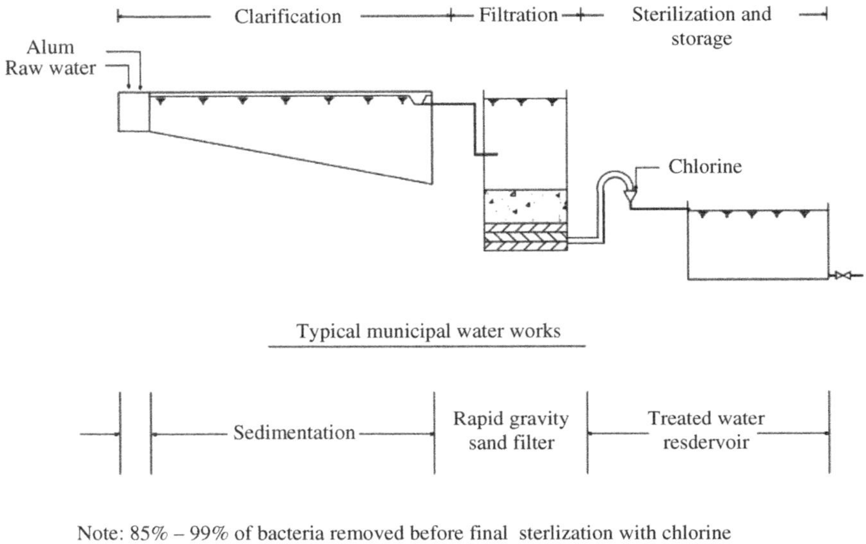

Figure 6.6 Proposed river water treatment plant for MRC (Airy, 2005)

6.4.1.2 Wastewater at MRC

The MRC uses water flush toilets that use from 1.5 to 2 liters of water per flush. The wastewater from the site is routed to septic tanks; there is a septic tank for each building that has a flush toilet. The greywater from the kitchen and laundry rooms are routed to a stone soak-field. Figure 6.9 shows a site map of the MRC.

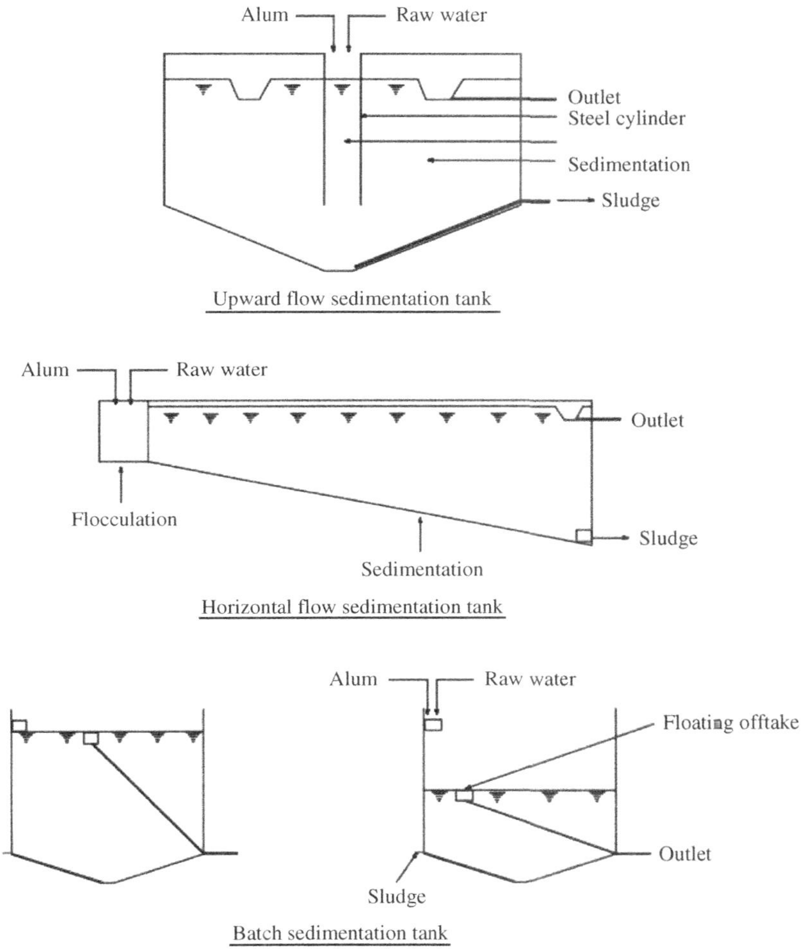

Figure 6.7 Three types of sedimentation tanks, proposal for MRC (Airy, 2005)

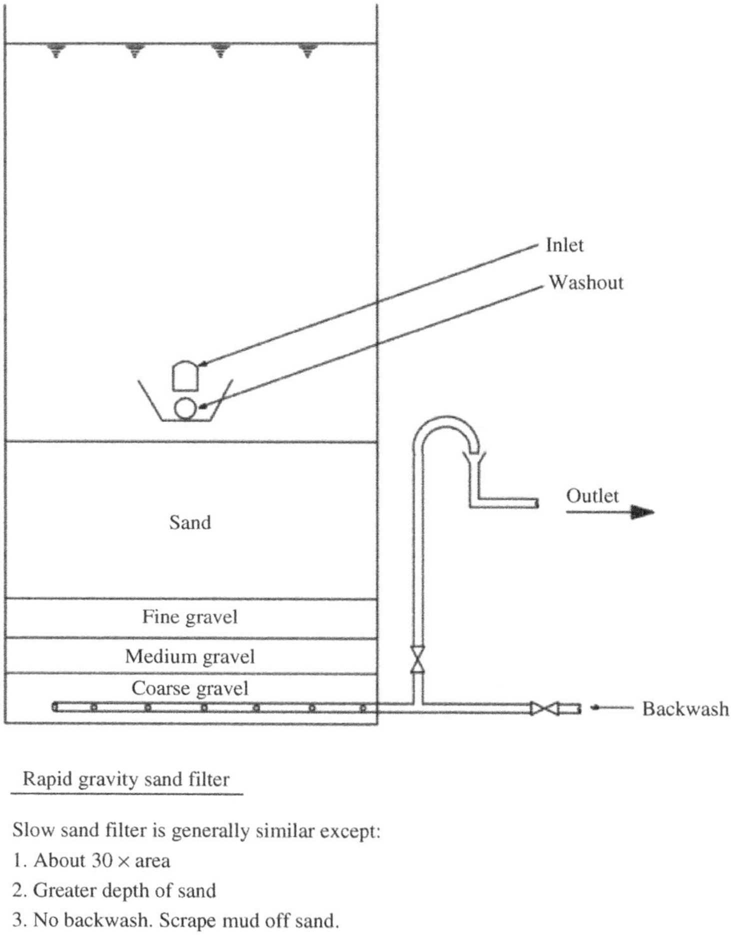

Figure 6.8 Rapid sand filter proposal for MRC (Airy, 2005)

6.4.2 Permanent village near MRC

The permanent village near the MRC is primarily inhabited by Kenyans who work in some capacity for the research center. The Ewaso Nyiro River water is used by the village for all water purposes. There are water faucets in the village that dispense water from the same 23,000 L river water tank that the MRC uses. Pit latrines are used for the containment of human waste.

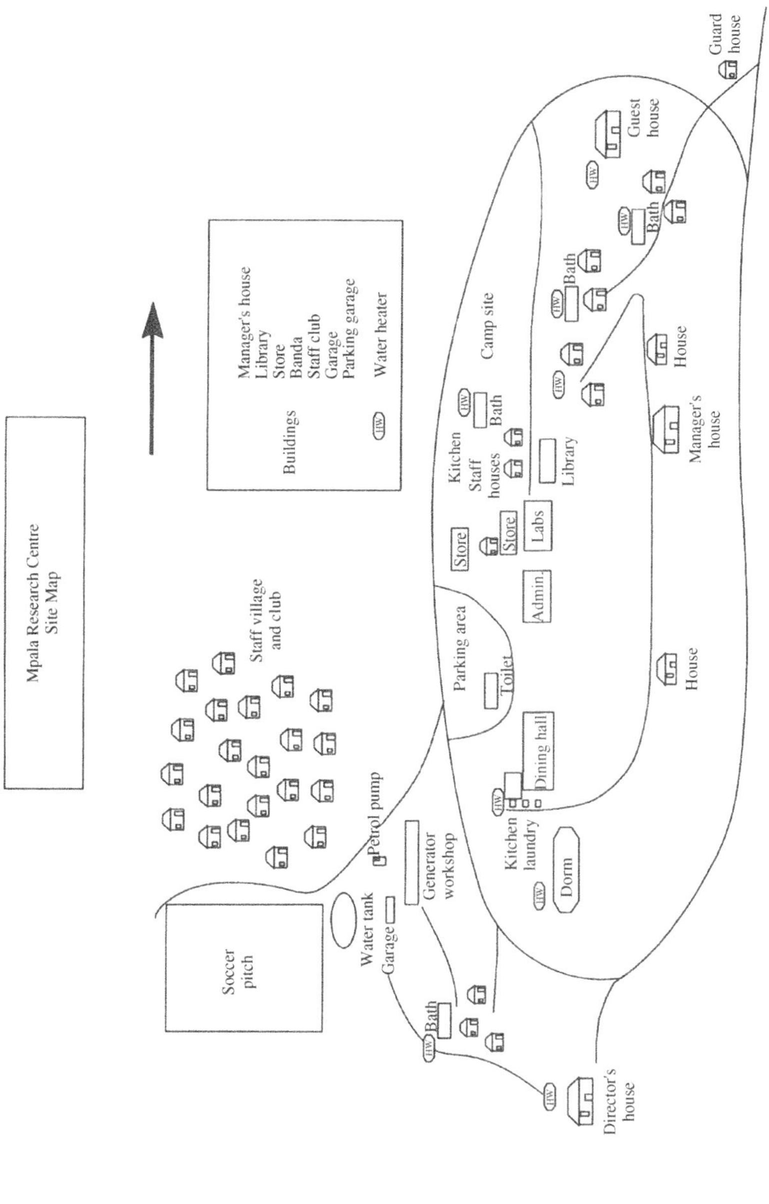

Figure 6.9 Site map of the Mpala Research Centre

There are 27 houses in the village, which are cylindrical structures made out of cement and stone with thatch roofs. According to Joe Petters, support services manager for MRC at the time of the study, the homes were built by Mpala and thatch was used for aesthetic purposes, even though it is generally cheaper to buy corrugated GI sheets for the roofing material (Petters, 2006). The actual cost of the thatch is not more expensive than the cost of iron sheets, but when shipping is factored into the costs, iron sheets are more economic. It costs roughly 4,000 Kshs to buy enough thatch to make two roofs, while it costs roughly 15,000 Kshs to buy enough iron sheets for two roofs (30 iron sheets per roof, 3 m × 1 m each, at 250 Kshs per sheet). It costs 14,000 Kshs to ship materials from Nanyuki to the MRC region. However, a truck load of thatch can only carry enough thatch for two houses, while one truck can carry enough iron sheets and supporting materials to construct roofs for an entire village. The circumference of a typical thatch roof in the village is 72 ft (about 22 m), which means the radius of the base of the roof is about 11.5 ft (3.49 m). The roof area (circular base) is about 413 sq. ft (38 sq. m).

I spoke with people in the village about water issues, rainwater harvesting, and composting toilets. The details of four conversations are given. The first person I talked with was Selina Ngasike. She felt that there was an inadequate quantity of water during the dry season and that it was also dirtier during that time. She and her community did not normally boil the water because of time constraints and the inconvenience it can bring. Also, she stated that it is difficult to tell when boiling is necessary, since sometimes people get sick from drinking the river water, but other times they do not. When asked about rainwater harvesting, Selina Ngasike preferred having a household rainwater harvesting system than having a community one. She noted that people often misuse the water when it is communal, and they may subsequently run out of water supply. She said that she preferred the corrugated, galvanized iron roofs, known as "mabati." When asked about latrines, called "choo" in Swahili, Selina Ngasike thought that the current pit latrines in the community were too "smelly." Of the various types of composting toilets, she preferred the Arbour Loo (AL) type, since one can make good use of the crop grown in the pit.

The second person interviewed was Sumat Sumbale. He stated that somehow, there is enough water for the community. However, because the river is so low during the dry season, they have to make a small dam with sand bags to help direct the water so that it can be pumped uphill to the river water storage tank. Sumat Sumbale explained that the quality of the water was not always good and that they have had people become sick with typhoid and dysentery. He preferred the mabati roofs to the thatch because the mabati is good for rainwater harvesting and it is longer-lasting. Like Selina Ngasike, Sumbale preferred that each household harvest its own rainwater as opposed to a communal system because he believed it would help to conserve water. When asked about latrines, Sumbale responded that the present pit latrines attract insects and create unpleasant odors. Of the various types of composting toilets, Sumbale preferred the AL type because he was not comfortable with transporting human waste; he would rather plant seeds in a pit than dig it up to be used as fertilizer.

The third person that I spoke with was Ramanto Manyas. Manyas stated that water is stable throughout the year, but water quality is poor because it is polluted by wildlife and humans. He noted that the people in his community sometimes experience health problems, such as cholera, when they drink the water. Like both Ngasike and Sumbale, Manyas prefers the mabati roofs because they stay cleaner than thatch, are long lasting, provide good shading, and are more stable against the wind and rain. He also stated that it is important that each household collect its own rainwater to avoid conflicts over the water resource. Manyas did not have any major problems with the current pit latrines, but he said that he would prefer the Fossa Alterna (FA) type of composting toilet because by alternating between two pits, he could avoid digging many pits around the community. He stated that the FA type of toilet would make good fertilizer for crops, and that he would be willing to handle the composted human waste.

The fourth person that I spoke with in the permanent settlement near the MRC was Patrick Etelej. He stated that there was not a sufficient quantity of water, especially during the dry season. During the rainy season, there are nearby streams with water, but these streams dry up during the dry season. Etelej affirmed that the water quality was not adequate for drinking, and he explained that drilling boreholes may be a solution to providing water for those who live far from the river. He preferred mabati roofs to thatch roofs because they are long-lasting and are good for rainwater collection. He also explained that there is a difference between the thatch that the MRC uses for their buildings and to build the permanent settlement roofs and those used by most villages in the area. Most communities use kangaroo grass, known as Themeda triandra, which is supposed to last six months. However, the MRC uses a more expensive and longer lasting thatch, mainly Pennisetum stramineum and Pennisetum mezianum. Etelej preferred the idea of household rainwater collection so that each household could take care of its own tank and roof, and determine how clean to keep it and how much water to draw from it. When asked about latrines, Etelej preferred both the AL and FA types of toilets to the current pit latrines, but he liked the AL type best because it did not involve as much handling of human waste.

6.4.3 Il Motiok

Three places were explored in Il Motiok—a new eco-tourism lodge, an old primary school, and the new Il Motiok Primary School and nearby village.

The eco-tourism lodge is funded by the Community Development Trust Fund (CDTF), a joint initiative of the Government of Kenya and the European Commission with the objective to contribute to poverty alleviation, as a biodiversity conservation project (BCP). The eco-tourism lodge in Il Motiok consisted of many circular thatch-roof lodges (Figure 6.10). The thatch is made of reeds that are gathered from water that is about 80 km to 100 km from the site. The plan is to lease the lodge to an investor for a set period of time, and a portion of the profits will go to the nearby community. Although there are no RWH plans for the thatch lodges, a total of 100 square meters of roof area will collect rainwater. These roofs are corrugated GI material and are for buildings such as the kitchen and staff offices.

Corrugated GI Roof of Kitchen for RWH | Thatch Roof of Common Area for Lodge Guests

Figure 6.10 Eco-tourism lodge in Il Motiok

On the visit to Il Motiok, I observed an old primary school which is now a storage place for construction materials (Figure 6.11). The school was about seven years old, and it was made out of wood with a corrugated GI roof. The roof was guttered along one side and had a neighboring ferrocement tank for rainwater storage. The tank was cracked; ferrocement tanks fail in this way if they are left dry for too long. The area of the base of the roof of the old primary school was 33 ft × 18 ft (10 m × 5.5 m).

Figure 6.11 The old Il Motiok Primary School and rainwater harvesting ferrocement tank (Base of roof dimensions: 33 ft $\times$18 ft; 10 m$\times$5.5 m)

The new Il Motiok Primary School started in 2000 and had completed buildings in 2002. It hosts about 86 students and has a plastic 6,000 liter RWH tank which collects water from one side of a corrugated GI roof (Figure 6.12). The dimensions for the base of the roof are 50.5 ft$\times$30 ft (15.4 m$\times$9 m). A non-governmental organization called SARDEP helped to purchase the GI sheets for the roof and rainwater tank.

I spoke with a teacher at Il Motiok Primary School, Agnes Keshine, about water issues, rainwater harvesting, and composting toilets for both the school buildings and the homes in her village. She acknowledged the lack of water to meet the demand at the school and in her village; also, Keshine noted that the river water and stored rainwater was not of good quality for both the community where she lives and the school in which she teaches. She believes that chlorine should be added to both the rainwater tank at the primary school and the river water for the community. Most of the houses in her community have thatch roofs because it is locally available; only two of them have iron roofs. Some houses have the traditional cow dung as the roof; this type of roof is becoming unpopular because of the smoke and dust problems created when food is cooked in the home. The semi-nomadic Massai to which she belongs have not been able to move as often as normal because of the drought in Kenya and subsequent lack of available grass. Keshine said that she prefers the GI sheets for roofs because there are no leaks, they can help collect rainwater, and they are long lasting. Also, she noted that thatch can be a fire hazard. When asked about the pit latrines at the school, Keshine was not satisfied with the odors and said that it would be better if they had a separate latrine for boys and girls. Figures 6.13 and 6.14 show homes in Il Motiok.

Gutter on one side of school

Plastic RWH tank

Lock on RWH tank

Figure 6.12 ll Motiok Primary School and rainwater harvesting tank (Base of roof dimensions: 50.5 ft×30 ft; 15.4 m×9 m)

Massai homes with thatch roofs

Massai house with cow dung roof

Figure 6.13 Massai homes (Bomas) in Il Motiok

6.4.4 Lekiji

In the village of Lekiji, Chief Isaac Hassan Lenguyo showed and discussed the various school buildings, homes, pit latrines, gardens, and local water resources. Lekiji has about 120 houses that are all constructed with various roof materials; these materials include thatch, corrugated GI, and earthen materials (refer to Figure 6.15a and 6.15b for the mixture of roof types). The base of the roof of the chief's home is 28 ft × 17 ft (8.5 m × 5 m) (Figure 6.16). Although there is no RWH currently taking place there, the chief plans on guttering one side of the house and connecting the gutters to one storage tank.

There are three classroom buildings and one teacher home building for Lekiji's primary school system (Figure 6.17); Figure 6.18 shows a diagram of the four buildings. According to Nelson Kiprop Murgor, a school teacher in Lekiji, there is not enough water, nor is there good quality water (due to lack of treatment) for the school children. Rainwater is pumped from the Nanyuki River to a 6,000 liter plastic storage tank near the school. The pump and tank were donations; the community pays for the fuel for the pump. The four buildings in the Lekiji School quarters all have corrugated GI roofs (Figure 6.18). Both the chief of Lekiji and Murgor pointed to cost of gutters and storage tanks as being major inhibitors to installing a RWH system.

Figure 6.14 House in Il Motiok with corrugated GI roof (Base of roof dimensions: 21.7 ft × 15.6 ft; 6.6 m × 4.8 m)

Unfinished House in Lekiji (Base of Roof Dimensions: 16.7 ft x 9.8 ft; 5 m x 3 m)

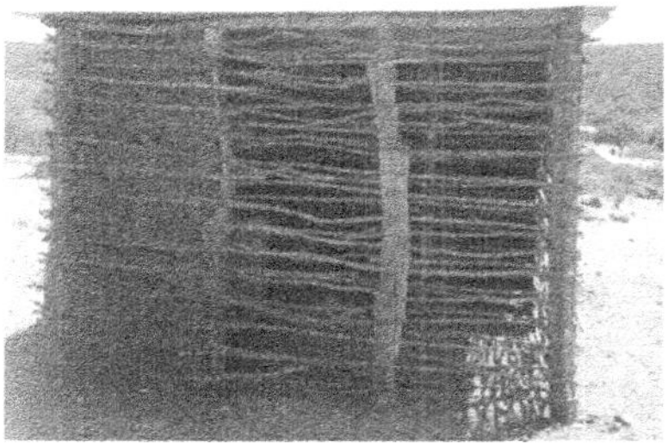

Unfinished House in Lekiji (Base of Roof Dimensions: 16.7 ft x 9.8 ft; 5 m x 3 m)

Figure 6.15a A variety of homes and roof types in Lekiji

Figure 6.15b A variety of homes and roof types in Lekiji

Side of house Roof of house

Figure 6.16 The home of the chief of Lekiji

Teacher homes

River water storage tank

School kitchen

School office

Pit latrine for teachers

Roof of Classroom b

Classroom a

Classroom c

Figure 6.17 Lekiji school buildings

Figure 6.18 is a layout representation of the three classrooms (a, b, and c) and the teachers' house (d) at Lekiji Primary School. All of the structures are made out of corrugated GI roofs with mud and stick bases. The dimensions of the base of the roof for **building a** is **32.5 ft × 14.6 ft (9.9 m × 4.4 m)**; the dimensions of the base of the roof for **building b** is **54.5 ft × 14 ft (16.6 m × 4.3 m)**; the dimensions of the base of the roof for **building c** is **61.5 ft × 16 ft (18.7 m × 4.9 m)**; the dimensions of the base of the roof for **building d** is **42 ft × 15.3 ft (12.8 m × 4.7 m)**.

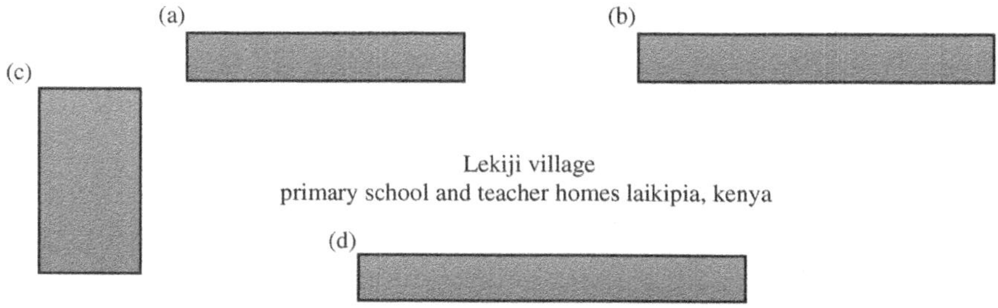

Figure 6.18 Lekiji village Primary School and teacher homes

With regard to latrine use, neither the Chief nor Murgor were satisfied with the pit latrines at the school because each year they were required to dig new pits. Both current and abandoned pits have odors (Figure 6.19). They preferred the FA-type of composting toilet because it eliminated the constant digging of pits and could provide useful fertilizer for their crops. Lekiji has a 1.5 acre (about 6,000 sq. m) community garden that currently harvests cabbage, tomatoes, and kales; rainwater is pumped uphill for irrigation (Figure 6.20). The water is pumped for a total of 4 hours each day (from 9 a.m. to 11 a.m. and from 3 p.m. to 5 p.m.). There is also a 1.5 acre (about 6,000 sq. m) private garden for one of the residents of Lekiji; river water is also pumped uphill to irrigate this garden and onions are grown in addition to the vegetables grown in the community garden (Figure 6.21). The chief of Lekiji had a suggestion: use the river water to irrigate and use the rainwater for drinking, bathing, and cooking purposes.

Pit latrine in for school pupils

Old pit latrine location

Figure 6.19 Latrines in Lekiji

Figure 6.20 Lekiji community garden (cabbage, tomatoes, kales)

Figure 6.21 Private garden in Lekiji (tomatoes, cabbage, kales, onions)

6.5 CONCLUSION

The quantity and quality of water available to MRC and the nearby communities is of grave concern. The administration of MRC has sponsored studies that analyze water supply, water consumption, and the possibilities for a sustainable water source for the center. Currently, MRC collects rainwater from 13 buildings; the rainwater is boiled and ceramic-filtered before it is used for drinking. Assuming a roof runoff coefficient of 0.9 (see Chapter 8), about 1 million liters of rainwater can be harvested each year from those buildings (Odhiambo, Petters and Wreford, 2005). Although this is not enough water to meet all of the water needs at the MRC, in an average year, water for drinking and cooking can be provided for the MRC population for most months (Odhiambo, Petters and Wreford, 2005). An independent study of water resources at MRC indicated that the river water supply can become more sustainable if (1) the river water pumps are improved and (2) a treatment plant is installed to clean the river water.

The findings from speaking with individuals at MRC, at the permanent village near MRC, in Il Motiok, and in Lekiji, show that there are similarities and differences among people's perceptions about the water aspects of sustainable buildings. Most people felt that there was inadequate water quantity and quality for their needs and that composting toilets would be a good alternative to the presently-used pit latrines. Some people preferred the AL type of toilet while others preferred the FA type. It is not straightforward whether or not roofing material is a barrier to harvesting rainwater in Laikipia. From my interviews, I found there to be no cultural barrier to using metal roofs instead of thatch in order to collect a better quantity and quality of rainwater from a roof. However, in some situations the thatch may be less expensive than the corrugated GI roofs. Thatch roofs have the advantage of being cooler than iron roofs in the hot sun. However, the corrugated GI roofs are better for collecting a large quantity and good quality of water; the runoff coefficient for thatch has been estimated at 0.2, while the runoff coefficient for GI is estimated at 0.9 (see Chapter 8). The thatch collects more particulate matter, is subject to termites, and does not last as long as the GI roofs. One issue that is worth pointing out is the importance of aesthetics to the eco-tourism lodge and nearby areas. Thatch is generally considered more aesthetically-pleasing than the GI roofs. The policy considerations of this tradeoff will be discussed in Chapter 9, Policy Considerations and Conclusions.

7
Water Quality Testing at Kenya Site
The village homes near the Mpala Research Centre Laikipia, Kenya

7.1 INTRODUCTION: WATER QUALITY TESTING AT KENYA SITE

This chapter details the experimental set-up, procedures, and results of the water quality experiments completed in Laikipia, Kenya. The following parameters were tested for rainwater and river water samples in Laikipia: turbidity, total and fecal coliform, sulfates, pH, color, nitrates, and nitrites. Water samples were taken from five sources: the Ewaso Nyiro River (1 km from MRC, in Il Motiok, and 15 km from MRC), a river faucet near the MRC, a rainwater storage tank at MRC, a rainwater storage tank at Il Motiok Primary School, and the Nanyuki River in the village of Lekiji. In addition, data from a water quality study of the

©2010 IWA Publishing. *Sustainable Water Resources in the Built Environment*. By Marilyn Waite. ISBN: 9781843393238. Published by IWA Publishing, London, UK.

Ewaso Nyiro River and Nanyuki River by the Ministry of the Environment and Natural Resources of Kenya is used as a supplement to this study.

7.2 RAINWATER HARVESTING WATER QUALITY EXPERIMENT

7.2.1 Sampling sites and sample collection

The following water locations were tested in the experiments:

(1) Ewaso Nyiro River (Figures 7.1, 7.2 and 7.3)

Figure 7.1 Trial 1, river approximately 1 km from Mpala

Figure 7.2 Trial 2, river in Il Motiok with cattle

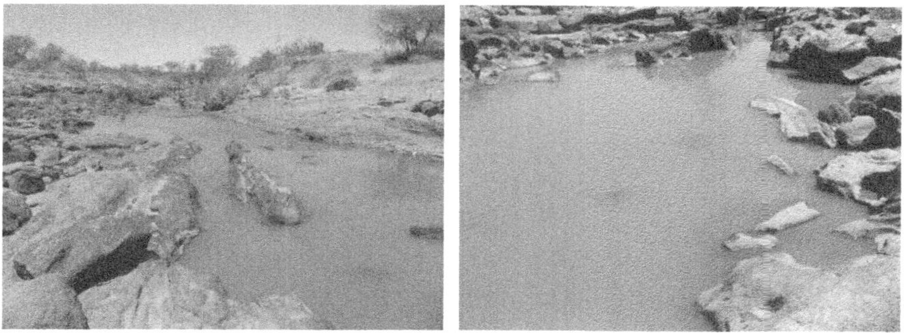

Figure 7.3 Trial 3, river approximately 15 km from Mpala

(2) River faucet near the MRC garage (Figure 7.4)

Figure 7.4 Trial 1

(3) Rainwater Storage Tank – MRC (Figure 7.5)

Figure 7.5 Trial 1, MRC rainwater storage ferrocement tank (from corrugated galvanized iron roof)

(4) Rainwater storage tank – Il Motiok Primary School (Figure 7.6)

Figure 7.6 Trial 1: Il Motiok Primary School rainwater storage plastic tank (from corrugated galvanized iron roof)

(5) Nanyuki River in Lekiji Village (Figure 7.7)

Figure 7.7 Trial 1: Sunday, January 29, 2006

7.2.2 Materials

The following materials were used to complete the experimental methods for all tests: 1,000 mL flasks, beakers, and bottles, 10 mL and 25 mL test tubes, 25 mL and 50 mL graduated cylinder, goggles, clippers, tape, and a permanent marker. All samples were stored at 1–2°C if they could not be analyzed immediately. To complete the pH test, a Baker-pHix wide ranging (pH 0.0–14) pH indicator stick kit was used. The following materials were used in the coliform tests: Coliscan Easygel® Easygel®, 20 mL bottles, pretreated petri dishes, 1 mL pipeter and pipet tips, ColiQuant EZ Manual, ColiQuant EZ Colony Color Guide, and an incubator. For the colorimeter tests, the following were utilized: Sulfate SulfaVer® 4 Reagent Powder Pillows, Nitrate HR NitraVer® 5 Reagent, Nitrogen-Nitrite HR NitriVer® 2 Reagent, Hach DR/890 Colorimeter, DR/890 Datalogging Colorimeter Handbook, and Kimwipes®.

7.2.3 Methods

The methods employed for the Kenya case study water quality tests can be found in the **Chapter 5, Water Quality Testing at Panama Site** Methods Section. The only difference between the methods completed in the Panama case study and the methods employed in the Kenya case study is that the total suspended solids (TSS) test and pH test using the LaMotte octet comparator were not completed in Kenya due to lab equipment constraints.

7.3 RAINWATER HARVESTING EXPERIMENT RESULTS

7.3.1 Ministry of the Environment and Natural Resources of Kenya study

Tables 7.1 and 7.2 below present the bacteriological water testing results for the Ewaso Nyiro and Nanyuki River in Laikipia, as well as the physical/chemical water testing results for the Ewaso Nyiro River completed by the Ministry of the Environment and Natural Resources of Kenya. This data indicates high color (true color), turbidity, and iron values for the Ewaso Nyiro River, as well as very high contamination with general, E. coli, and fecal coliforms for both the Ewaso Nyiro and Nanyuki rivers.

Table 7.1 Results of bacteriological examination of water in Laikipia completed on February 22, 2005 by the Republic of Kenya's Ministry of the Environment and Natural Resources (Ministry of Water and Irrigation Laikipia District, 2005)

Source of Sample: Ewaso Nyiro River near pumping station at MRC
Exact Site Sample taken from: The middle of the river

Total Coliforms	11,000 CFU/100 mL
E.Coli	110 CFU/100 mL
Fecal Coliforms	4,600 CFU/100 mL

Source of Sample: Nanyuki River
Exact Site Sample taken from: The middle of the river

Total Coliforms	11,000 CFU/100 mL
E.Coli	40 CFU/100 mL
Fecal Coliforms	2,100 CFU/100 mL

7.3.2 Coliform results

Table 7.3 shows the fecal coliform and total coliform count for water from (1) the Ewaso Nyiro River 1 km from MRC, (2) a faucet near the MRC garage that dispenses the Ewaso Nyiro River water that has been pumped uphill into a 23,000 L tank, (3) the rainwater stored in a ferrocement tank at MRC, (4) the rainwater stored in the plastic tank at the Il Motiok Primary School, and (5) the Nanyuki River in the village of Lekiji. The coliform results in Table 7.3 show that all sources tested indicate the presence of general coliforms (see the Methods Section of **Chapter 5** for the distinction between total, fecal, and general coliforms). The only site that did not have any fecal coliforms was the plastic rainwater storage tank at Il Motiok Primary School. The most fecal and total coliforms were found from the Ewaso Nyiro River sample taken by the Ministry of Environment and Natural Resources of Kenya. The Nanyuki River Sample taken by the Ministry of Environment and Natural Resources of Kenya had the same amount of total coliforms as the Ewaso Nyiro River sample. Interestingly, the second-highest fecal coliform count was the rainwater storage tank at MRC.

Table 7.2 Physical/chemical water analysis results of the Ewaso Nyiro river near MRC Completed on February 22, 2005 by the Republic of Kenya's Ministry of the Environment and Natural Resources (Ministry of Water and Irrigation Laikipia District, 2005)

Parameters	Unit	Results
pH	pH Scale	8.0
Colour	mgPt/l	70
Turbidity	N.T.U.	40.25
Permanganate Value (20 min boiling)	mgO2/l	7.11
Conductivity (25°C)	μS/cm	338
Iron	mgFe/l	1.912
Manganese	mgMn/l	0.18
Calcium	mgCa/l	22.4
Magnesium	mgMg/l	8.76
Sodium	mgNa/l	36.25
Potassium	mgK/l	5
Total Hardness	mgCaCO3/l	92
Total Alkalinity	mgCaCO3/l	130
Chloride	mgCl/l	24
Fluoride	mgF/l	0.48
Nitrate	mgN/l	2.6
Nitrite	mgN/l	0.06
Ammonia	mgN/l	–
Total Nitrogen	mgN/l	–
Sulfate	mgSO4/l	<0.3
Orthophosphate	mgP/l	–
Total Suspended Solids	mg/l	–
Free Carbon Dioxide	mgCO2/l	4
Dissolved Oxygen	mgO2/l	–
TDS	mg/l	209
Silica	mgSiO2/l	

The high coliform count in the MRC RWH tank (unlike the Il Motiok Primary School RWH tank) could be attributed to the age of the tanks, the need for frequent cleaning of the tanks, and the abundant amount of wildlife at the MRC that contributes to fecal contamination. A likely source of fecal contamination is from pigeons that spend time on the rooftops; their feces are drained into the roof gutters and finally settle in the water tanks. The WHO and EC guideline for coliforms is 0 CFU/100 mL. The USEPA guideline for coliforms is less than 1 CFU/100 mL. In comparison to the WHO, EC, and USEPA standards for coliform, the rainwater collected from the MRC building and the Il Motiok Primary School building were not suitable for drinking.

Table 7.3 Total and fecal coliform results for Laikipia, Kenya study, January 2006

Source		Fecal Coliform (CFU/100 mL)	Total Coliform (CFU/100 mL)
Ewaso Nyiro River (near MRC) ◇	**Trial 1**		
		4,600	11,000
Ewaso Nyiro River (MRC Faucet)	**Trial 1**		
	Petri Dish 1	60	620
MRC Rainwater Storage Tank	**Trial 1**		
	Petri Dish 1	2,900	7,900
Il Motiok Primary School Rainwater Storage Tank	**Trial 1**		
	Petri Dish 1	0	80
Nanyuki River	**Trial 1**		
(Lekiji)	Petri Dish 1	180	1,120
	Petri Dish 2	200	1,200
Nanyuki River ◇	**Trial 1**		
		2,100	11,000

Note: CFU: Colony Forming Units, Amount of Sample Used (mL): 5, ◇ indicates that these results were obtained by the Ministry of Environment and Natural Resources of Kenya during the dry season on February 22, 2005. All other results were obtained by my experiment during the dry season in late January 2006.

7.3.3 Results for all other parameters

Table 7.4 shows results for the four water locations tested for the following water quality parameters: pH, turbidity, sulfates, nitrates, nitrites, and apparent color. The sulfate, nitrate, and pH values were within the acceptable limits of WHO, USEPA, and EC drinking water standards. However, some of the nitrite, turbidity, and color values were not within the standards. The nitrite values for the Nanyuki River and the Il Motiok Primary School rainwater storage tank exceeded the standards set by WHO, USEPA, and the EC. All of the water samples also exceeded turbidity standards set by the three organizations (although the turbidity values found in the RWH tanks at MRC and Il Motiok may be outside of the detection limit for the Hach test). The color values for all of the water samples vastly exceeded the standards set by the WHO and EC (USEPA does not provide a standard); however, it is worth noting that the values obtained in the Kenya experiment are only for apparent color, while the WHO and EC standards are set for true color. It can be reasonably assumed that the true color of these samples may exceed the standards since the apparent color values are very high. The fact that the nitrite, turbidity, and color values exceeded the limits set by the WHO, USEPA, and EC, further indicates that the rainwater should be treated before it is used as a potable water source.

Table 7.4 Results for all other parameters tested (pH, turbidity, sulfates, nitrates, nitrites, and apparent color) in Laikipia, Kenya experiment, January 2006

Parameter	Source					
	Ewaso Nyiro River (near MRC, in Il Motiok, far MRC)	Ewaso Nyiro River ◇ (near MRC)	Ewaso Nyiro River Faucet (MRC)	Rainwater Storage Tank (MRC)	Rainwater Storage Tank (Il Motiok)	Nanyuki River (Lekiji)
ph (Litmus test)						
Trial 1	7	8	7	7	7	7
Trial 2	7	–	–	–	–	–
Trial 3	7	–	–	–		–
Turbidity (FAU) [EDL: 21]						
Trial 1	118	40.25 (NTU)	43	9 (ND)	6 (ND)	21
Trial 2	44	–	–	–	–	–
Trial 3	81	–	–	–	–	–
Sulfates (mg/L) [EDL: 4.9]						
Trial 1	2 (ND)	< 0.3	5	0 (ND)	0 (ND)	0 (ND)
Trial 2	1 (ND)	–	–	–	–	–
Trial 3	0 (ND)	–	–	–	–	–
Nitrates-Nitrogen (mg/L) [EDL: 0.8]						
Trial 1	0 (ND)	2.6	0.3 (ND)	0.9	0 (ND)	1.6
Trial 2	0 (ND)	–	–	–	–	–
Trial 3	0 (ND)	–	–	–	–	–
Nitrites-Nitrogen (mg/L) [EDL: 2]						
Trial 1	0 (ND)	0.06	0 (ND)	0 (ND)	4	5
Trial 2	0 (ND)	–	–	–	–	–
Trial 3	0 (ND)	–	–	–	–	–
Color, Apparent (Pt-Co Units) [EDL: 25]					–	
Trial 1	550	70 (True Color)	355	62	123	212
Trial 2	463	–	–	–	–	–
Trial 3	550	–	–	–	–	–

Note: – indicates Not Tested; ND: Non-Detectable; EDL: Estimated Detection Limit; ◇ indicates that these results were obtained by the Ministry of Environment and Natural Resources of Kenya during the dry season on February 22, 2005. All other results were obtained by my experiment during the dry season in late January 2006. 1 NTU = 1 FTU = 1 FAU for the Hach test used in this experiment.

7.4 SUMMARY

The water quality tests performed in Laikipia, Kenya from this experiment and by the Ministry of Environment and Natural Resources of Kenya indicate that the Ewaso Nyiro and Nanyuki rivers, as well as rainwater sources from MRC and Il Motiok Primary School, are heavily contaminated with coliforms. In addition, the water samples had turbidity values that exceeded the recommended value set by the WHO. One of the key lessons learned from this experiment regarding RWH is the need for rainwater treatment for potable uses. In comparison to the WHO, EC, and USEPA standards for coliform, the rainwater collected from the MRC building and the Il Motiok Primary School building are not suitable for drinking. It is likely that by diverting the roof runoff from the first few minutes of a downpour, the contaminants on the roofs (such as bird droppings) will not be washed into the tank.

8
Rainwater harvesting design
The village homes near the Mpala Research Centre Laikipia, Kenya

8.1 INTRODUCTION: RAINWATER HARVESTING DESIGN

One goal of the Kenyan case study is to help design a rainwater harvesting system for various buildings in Laikipia. This chapter presents information relevant to designing RWH systems in general, as well as specific design recommendations for buildings analyzed in the case study. In particular, a discussion of the following is provided: RWH classification, roof catchments, storage tank volume, building materials for the storage tank and gutter, RWH system efficiency, and aspects of gutter design including water flow, interception, mounting, and pipes for connecting the gutter to the storage tank.

ISBN: 9781843393238. Published by IWA Publishing, London, UK.

In addition, a discussion of bacterial die-off in the storage tank and technologies that prevent and treat the contamination of rainwater (including filters, first flush devices, fine filtering, chlorination, and solar treatment) are provided.

The UNEP has outlined several key elements of designing a rainwater collection system that is similar to the approach taken in this chapter. Some of the steps needed to plan a successful RWH system include determining the demand for rainwater through surveys and questionnaires, including a procedure where water rationing begins, determining the types of building materials and buildings sizes, incorporating the concept of the first flush using a device that is easy to operate, and ensuring guttering is simple and adequate to collect rainwater from a sufficient area relative to tank size (UNEP, 1999). According to UNEP, it is good practice to standardize materials at the village level for maintenance purposes. Also, storage tank volumes should be based on rainfall analyses and predicted water usage.

8.2 DESIGNING A RAINWATER HARVESTING SYSTEM

8.2.1 Classification

RWH near Mpala can be used as an intermittent, partial, or full source for water. The Development Technology Unit (DTU) at the University of Warwick has described these three types of sources as follows. *Intermittent* describes the situation in which the requirements of the user are met for a part of the year. During the dry season, an alternative water source has to be used or water is carried in from local surface water and stored in the RWH tank. RWH used as a *partial* source provides limited coverage of the user's water requirements during the entire year. For example, a family would collect rainwater to meet their high water-quality needs such as drinking, while other needs such as bathing are met by another water source. This can be achieved in an area with a single or two short wet seasons and a large storage capacity to cover the needs during the dry season (DTU, 2005f). *Full* describes the situation in which the total demand of the user is met for the entire year by only rainwater. According to DTU, this type of system is attractive for areas that have a bimodal rainfall pattern because the tank will be recharged during both wet seasons. *The RWH system in Laikipia can most reasonably be designed for partial use throughout the year given the drought conditions, cost barriers to purchasing large storage tanks, and the limited size of the roof catchment areas.*

8.2.2 Roof catchments

When designing a RWH system, it is important to consider the characteristics of the catchment surface. The catchment determines the quantity and affects the quality of water that enters the storage tank. Most losses of rainwater occur through infiltration, but some water may also be lost as it bounces off the edge of the surface in heavy downpours (Thomas and Martinson, 2007). This loss is represented by a run-off coefficient, C_R, which is a number between zero and one. The quantity $(1-C_R)$ expresses the loss fraction averaged over a year. Thus, a good impermeable roof will have a C_R value close to 1, while a less impermeable roof will have a C_R value close to 0. In Table 8.1, various roof types with corresponding runoff coefficients and expected water quality are given.

Table 8.1 Characteristics of roof types (Thomas and Martinson, 2007)

Type	Runoff coefficient	Comments
GI Sheets	>0.9	• Excellent water quality. Surface is smooth and high temperatures help to sterilize bacteria.
Tile (glazed)	0.6–0.9	• Good water quality from glazed tiles. • Unglazed tiles can harbor mould. • Contamination can exist in tile joins.
Asbestos Sheets	0.8–0.9	• New sheets provide good water quality. • No evidence of carcinogenic effects by ingestion. • Slightly porous, so reduced runoff coefficient and older roofs harbor mould.
Organic (Thatch, Cadjan)	0.2	• Poor water quality. • Little first flush effect (see Chapter 2, Reducing Contamination). • High turbidity due to dissolved organic material that does not settle.

The two primary roofing materials found in the Kenya case study were thatch and corrugated galvanized iron (GI) roofs. Based on the interviews conducted in Kenya, I found that there were no cultural barriers, and in some cases, no cost barriers to using the GI roofs for construction (see Chapter 6). Thatch roofs are known to provide poor water quality and water quantity for rainwater collection. Due to financial obstacles, thatch roofs are still sometimes used for rainwater collection in rural settings across the developing world. Various treatments for thatch roofs have been tried such as polythene sheeting, however the sheeting tends to degrade in the sunlight quickly and can only be used for a single season (Thomas and Martinson, 2007). One solution has been tried in Ethiopia in which a foldaway roof was constructed; however, problems still arise because the system relies on user intervention to fold and unfold the roof at height and when it is raining. According to DTU, a more practical solution may be the use of auxiliary catchments (Thomas and Martinson, 2007). Since the thatch roofs have a very low runoff coefficient (0.2) and will collect generally dirty water, it is not recommended that this roof material be used to harvest rainwater for potable purposes. The water may be good, however, for small-scale irrigation, in which case a configuration using a collection pan, wooden trough, and some type of elevation mechanisms such as a small ladder can be used to collect the water from the roof.

8.2.3 Storage tank volume

8.2.3.1 Rainwater supply from buildings averaged over a year

In order to design the size of the RWH tank, one can use two main approaches: design the tank based on the demand or design the tank based on the supply. The demand-side approach is appropriate for wet regions that have abundant rainfall; there is likely to be more rain than is needed and more rain than can be harvested in a cost-effective manner. The supply-side approach is appropriate for dry regions where rainfall may not be able to meet all of the water needs. Since Laikipia is a low rainfall area and the demand for water may not meet the rainfall supply, the supply-side approach for sizing the tank was used and compared with a pre-determined water demand. The equations used to calculate the amount of water that can be supplied are as follows (see Chapter 2):

$$\begin{aligned} Gross\ water\ supply &= (area\ of\ roof\ horizontal\ projection) \\ &\quad \times (annual\ rainfall) \end{aligned} \tag{8.1}$$

$$Net\ water\ supply = C_R \times (gross\ water\ supply) \tag{8.2}$$

$$Net\ water\ supply/number\ of\ days\ in\ a\ year = daily\ water\ storage\ volume \tag{8.3}$$

$$1\,m^3 = 1000\,liters\ (L) \tag{8.4}$$

where,

Gross water supply (yearly) is in m^3
Surface area is in m^2
Rainfall (yearly) is in m

The average annual rainfall for the MRC area ranges from roughly 400 mm (0.4 m) to 600 mm (0.6 m). The range of water supply volumes for various buildings near MRC is provided in Table 8.2 below using Equations (8.1)–(8.4).

Table 8.2 Potential water supply from different buildings in Laikipia using rainfall averaged over a year

Structure (6 Total)	Roof Projection Dimensions			Annual Net Supply	Annual Net Supply	Daily Net Supply	Daily Net Supply
	Length (m)	Width (m)	Roof Area (m^2)	(0.4 m of Rainfall) (L)	(0.6 m of Rainfall) (L)	(0.4 m of Rainfall) (L)	(0.6 m of Rainfall) (L)
Home in Il Motiok	6.6	4.8	31.68	11,405	17,107	31	47
Chief of Lekiji's House	8.5	5	42.5	15,300	22,950	42	63
Lekiji Primary School a	9.9	4.4	43.56	15,682	23,522	43	64
Lekiji Primary School b	16.6	4.3	71.38	25,697	38,545	70	106
Lekiji Primary School c	18.7	4.9	91.63	32,987	49,480	90	135
Lekiji Teacher Homes d	12.8	4.7	60.16	21,658	32,486	59	89

The United Nations estimates that the absolute daily minimum amount of water a person needs is 50 L (13.2 gallons): 5 L for drinking, 20 L for sanitation and hygiene, 15 L for bathing, and 10 L for preparing food (Kirby, 2002). Although alternative estimates for water needs exist (such as a total of 20 L of water per person per day) (UNESCO, 2003), the 50 L estimate will be used for consistency purposes; if water needs for drinking and other purposes are taken to be less, then one would expect to have more water available. *In this analysis, these water needs will be equated with water demand and water consumption* (*although it may often be the case that the water demanded is less than the need, especially in impoverished regions*). For an average household of five members, this yields a household water demand of 250 L per day for households in Laikipia. It is clear from the chart above that rainwater collected from the two homes in Il Motiok and Lekiji, as well as the teacher homes, would not be able to provide enough water for *all* domestic water needs (using rainfall data averaged over a year). *The domestic RWH can, however, provide enough water to meet the total household daily minimum drinking water needs of 25 L.*

The three Lekiji Primary School buildings would most likely want to provide enough water for drinking, cooking, and cleaning purposes in the school, which I estimate to be 6 liters per person per day (3 liters for drinking and 3 liters for cooking). Since there can be *roughly* 100 students and faculty at the school, the total daily demand for water is about 600 L per day. Since the amount of daily rainwater supply would only be from 203–305 L for all three school buildings, there would not be enough water to meet all water needs, but there would be enough water to meet the drinking water needs given the upper bound of rainfall (0.6 m of rain per year).

8.2.3.2 Designing storage tank size

Table 8.3 and Figure 8.2 show the monthly rainfall data in Laikipia, Kenya near MRC. Table A.11 in the Appendix shows detailed monthly rainfall data. The data was taken from two locations (the Mpala Conservancy and the Mpala Research Centre), using six full years of monthly rainfall data, from 1999 to 2004 (STRI, 2006). The mean monthly rainfall at the Mpala Conservancy is 45 mm with standard deviation of 31 mm, while the mean monthly rainfall at MRC is 49 mm with a standard deviation is 35 mm.

Table 8.3 Monthly rainfall averages at the Mpala Conservancy and MRC from 1999–2004 (STRI, 2006)

Month	Mpala Conservancy	MRC
	Rainfall Average (mm)	Rainfall Average (mm)
1	23	28
2	6	6
3	39	37
4	112	123
5	54	70
6	24	25
7	57	61
8	65	82
9	21	12
10	24	27
11	89	85
12	26	31

In order to design the rainwater tank sizes for the six various buildings near MRC, one must consider that the region is extremely drought-prone. Thus, monthly rainfall data may obscure peaks in rainfall that are worth taking into account in the design. The approach used for this design follows the philosophy that "every drop counts." Using daily rainfall data from the years 1999 to 2004, I analyzed the points in which there were consecutive days of notable rainfall. In practice, I took note of every time there was rainfall on two consecutive days or more. I then found the maximum rainfall for the consecutive days. I discovered that there were two instances of 15-day periods that had notable rainfall, with totals of 147 mm (November 2001) and 227 mm (April 2003) of rainfall each (see Table 8.4) (STRI, 2006). Interestingly, this outcome corresponds with the "short" and "long" rainy seasons found in general in Kenya. It also corresponds with the two months that have the highest average monthly rainfall (see Figure 8.1). In addition to designing for the more extreme rainfall events, one must also take into account that the cost of the tank will increase as the size of the tank increases. I therefore took the average of the two extreme sets of data instead of taking the highest total rainfall set from April of 2003. Table 8.4 and Figure 8.2 summarizes this data.

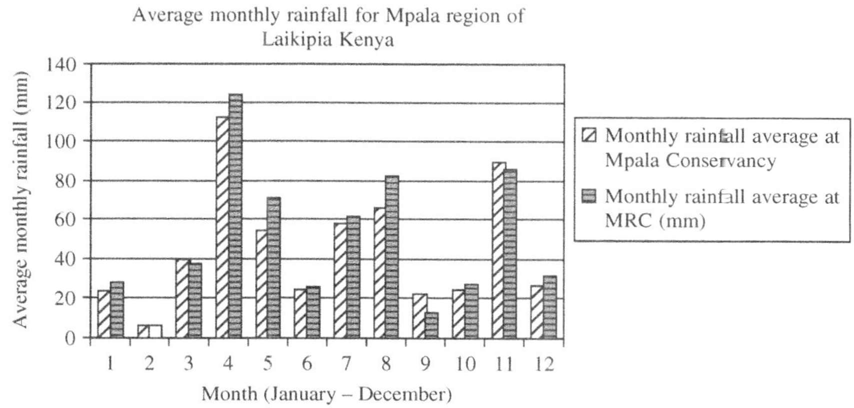

Figure 8.1 Average monthly rainfall at the Mpala Conservancy and MRC from 1999–2004 (STRI, 2006)

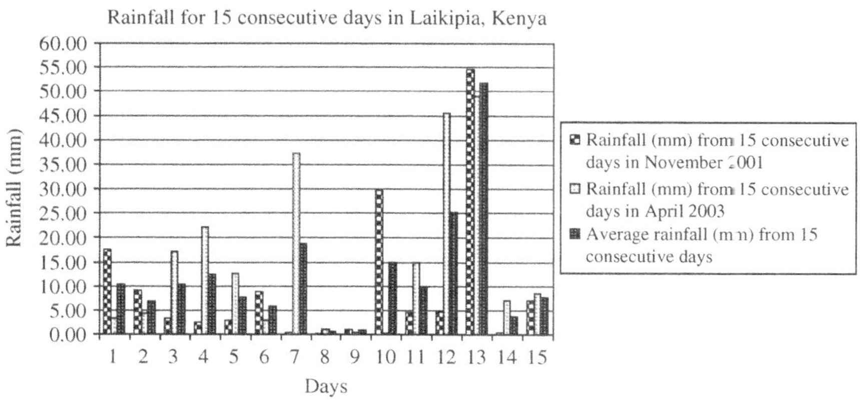

Figure 8.2 15-day consecutive rainfall data in Laikipia, Kenya for days in November 2001 and April 2003 (STRI, 2006)

Table 8.4 15-Day Consecutive Rainfall Data in Laikipia, Kenya for Days in November 2001 and April 2003 (STRI, 2006)

Day	Rainfall (mm) Nov-01	Apr-03	Average Rainfall for Day
1	17.53	3.30	10.42
2	9.14	4.32	6.73
3	3.30	17.27	10.29
4	2.54	22.10	12.32
5	2.79	12.70	7.75
6	8.89	2.79	5.84
7	0.51	37.34	18.92
8	0.25	1.02	0.64
9	1.02	0.51	0.76
10	29.72	0.25	14.99
11	4.57	14.99	9.78
12	4.83	45.47	25.15
13	54.61	49.02	51.82
14	0.51	7.11	3.81
15	7.11	8.38	7.75
Total Rainfall	147.32	226.57	186.95

Using the average rainfall values for the 15-day consecutive rainfall events, I then determined the storage volume, or tank size, by taking the *maximum* storage volume in the 15-day period. Storage volume was calculated by subtracting the daily water consumption from the daily water supply (found using concepts in Equations (8.1)–(8.3) and adding the amount of water stored from previous days to subsequent days (cumulative storage volume). Table A.12 in the Appendix show this method for all of the buildings analyzed in Laikipia. Water consumption was taken to be 50 L per person for the three households examined (house in Il Motiok, chief of Lekiji's house, and Lekiji teachers' houses) and 6 L per person for the students at school. It is assumed that out of the 100 people (estimate) at Lekiji Primary School, 15% would obtain water from building *a*, 25% would obtain water from building *b*, and 60% would obtain water from building *c*. This estimate is based on the size of the respective school buildings (see Chapter 6).

It is important to note that the size of the tank is dependent on three main variables: rainfall, roof catchment size, and water demand/consumption. Since the rainfall and roof catchment size are fixed for each building, the only flexible variable is how much water will be used from the tank each day. Thus, for this analysis, *the design size for the tank will change if the assumptions of daily water use are changed.* For example, if the water demand/consumption is limited to providing only drinking water for the six buildings in question, the design size for the tank will be bigger to account for the increase in storage needed for left-over water to be used in future days. For the three school buildings in Lekiji, the amount of water supplied for a certain amount of people was divided between the buildings. Alternatively, one could *not* design for a certain number of people, but rather design for a certain water need for the maximum number of people that can be serviced within that constraint.

Table 8.5 illustrates the method for determining an appropriate storage tank size using a home in Lekiji as an example. An example of how the storage volume is calculated each day can be seen by observing the first two rows of Table 8.5; the daily water demand of 0.25 m^3 is subtracted from the daily water supply of 0.40 m^3 to obtain the first storage volume of 0.15 m^3. Then the storage volume on day 2 is the storage volume from day 1 (0.15 m^3) added to the daily water demand of 0.25 m^3 subtracted from the daily water supply of 0.26 m^3; the result is a storage volume of 0.16 m^3. This type of sequence continues for the fifteen days, and the maximum storage volume in the fifteen day period becomes the design size for the tank.

Table 8.5 Calculation of storage volume for the chief of Lekiji's home (5 people [estimate] × 50 liters of water needed per day)

Day	Daily Water Demand (m^3)	Rainfall (m)	Roof Size (m^2)	Daily Water Supply (Roof Size∗Rainfall∗0.9)	Storage Volume (m^3)
1	**0.25**	0.0104	42.5	**0.40**	**0.15**
2	**0.25**	0.0067	42.5	**0.26**	**0.16**
3	0.25	0.0103	42.5	0.39	0.30
4	0.25	0.0123	42.5	0.47	0.52
5	0.25	0.0077	42.5	0.30	0.57
6	0.25	0.0058	42.5	0.22	0.54
7	0.25	0.0189	42.5	0.72	1.01
8	0.25	0.0006	42.5	0.02	0.79
9	0.25	0.0008	42.5	0.03	0.57
10	0.25	0.0150	42.5	0.57	0.89
11	0.25	0.0098	42.5	0.37	1.01
12	0.25	0.0251	42.5	0.96	1.73
13	0.25	0.0518	42.5	1.98	3.46
14	0.25	0.0038	42.5	0.15	3.35
15	0.25	0.0077	42.5	0.30	3.40
				Maximum Storage Volume (m^3)	**3.46**

Table 8.6 shows the recommended tank sizes for the various buildings near MRC. The tanks can be built out of various materials (discussed in next section), including ferrocement and plastic. Various prices for the materials needed to construct the tank, as well as plastic tanks in 460 L, 1000 L, 2,300 L, and 4,600 L sizes, were quoted from a local shop in Nanyuki, Kenya (see Figure A.10 in the Appendix). In a report by staff members at MRC, the cost to build one 13,000 L ferrocement tank (with stone) was estimated at 1,500 USD (Odhiambo, Petters, and Wreford, 2005).

Table 8.6 Recommended tank sizes for various buildings in Laikipia, Kenya

Structure	Recommended Tank Size (L) (rounded to nearest hundred)
House in Il Motiok	1,900
Chief of Lekiji's House	3,500
Lekiji Primary School a	6,000
Lekiji Primary School b	9,800
Lekiji Primary School c	10,000
Lekiji Teacher Homes d	4,300

An analysis showing the time period in which the storage volume reaches zero was completed. An example of the method is shown using the same house in Lekiji. I did two separate analyses using the data from the 2001 and 2003 daily rainfall periods. Starting with the average 15-day consecutive rainfall period for October 2001 and April 2003 (used in tank size determination, see Table 8.4), I continued to use the subsequent rainfall data for the remaining days of the 2001 and 2003 years. I thus performed two separate calculations—one based on the 2001 daily rainfall data starting after the fifteen-day consecutive rainfall period in November and one based on the 2003 daily rainfall data starting after the fifteen-day consecutive rainfall period in April (both starting with the average of the two 15-day consecutive rainfall events). Table A.13 in the Appendix shows the order of the rainfall data for these two separate calculations. I calculated the daily storage volumes with the extra rainfall data until the storage volume was zero or less than zero. Figures 8.3 and 8.4 illustrate the water demand, water supply, and storage volume for a home in Lekiji for the two time periods (October and April). These two time periods correspond to significant rainfall events in the short and long rainy seasons. For the home in

Lekiji, the analysis shows that there is adequate storage for roughly a month in November and roughly a month and a half in April that meet all domestic water needs (estimated at 50 L per person per day). The storage period would be longer if the rainwater were rationed, for example, to supply only drinking water. With the exception of Lekiji Primary School Buildings *a* and *b*, the storage volume generally lasts for a month. The Lekiji Primary School Building *a* and Building *b* have storage volumes that last for longer time periods (3 to 4 months) partly because the assumed daily demand is a lot less than the other buildings (900 L of water demanded per day for Building a and 150 L of water demanded per day for Building b). Figures A.11.a to A.11.l in the Appendix show graphs of water demand, water supply, and water storage volume for the six buildings in Laikipia. Table 8.7 summarizes the number of days after the 15-day consecutive rainfall period that it takes for the storage volume in the tank to be zero (completely empty) for six buildings in Laikipia.

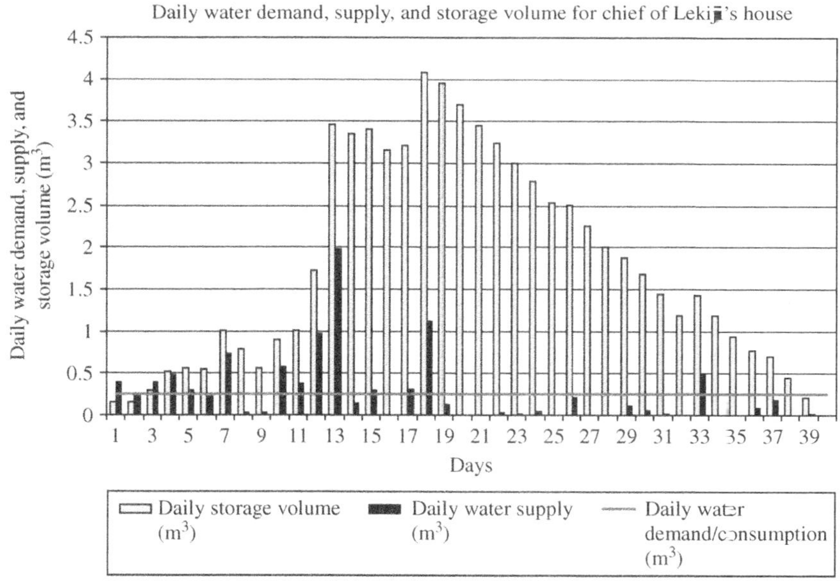

Figure 8.3 Daily water demand, supply, and storage volume for a house in Lekiji (Starting from significant rainfall event in *April 2003* and continuing until daily storage volume is zero or less)

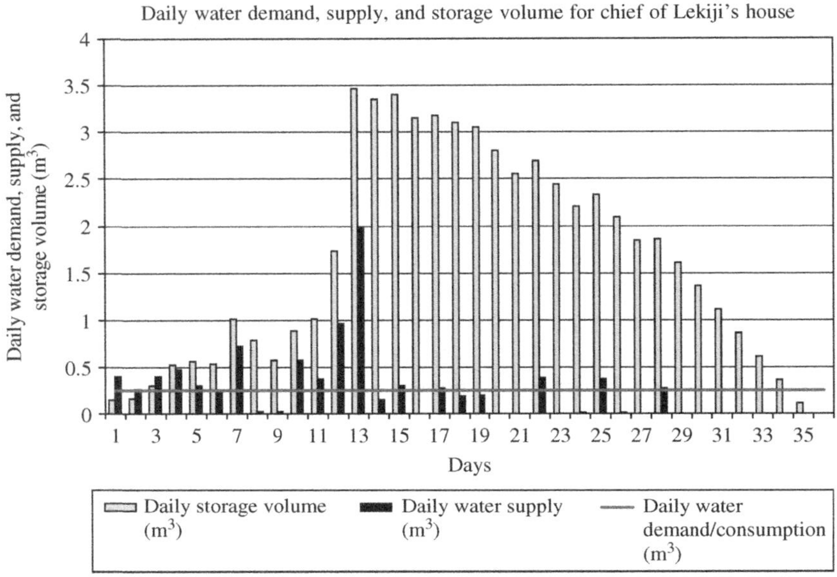

Figure 8.4 Daily water demand, supply, and storage volume for a house in Lekiji (Starting from significant rainfall event in *November 2001* and continuing until daily storage volume is zero or less)

Table 8.7 Number of days until tank volume is empty for six buildings in Laikipia

Structure	Number of Days Nov-01	Number of Days Apr-03
House in Il Motiok	12	13
Chief of Lekiji's House	21	25
Lekiji Primary School a	109	256
Lekiji Primary School b	106	251
Lekiji Primary School c	17	20
Lekiji Teacher Homes d	40	53

8.2.4 Building materials for storage tank and gutter

Commonly used materials for storage tanks include reinforced concrete, fiberglass, polyethylene, stainless steel, and ferrocement tanks and jars made of mortar or earthen materials. Storage tanks should be constructed of inert material. According to UNEP, gutters and conveyance pipes "should be made of plastic, PVC [Polyvinyl chloride] or other inert substance, as the pH of rainwater can be low (acidic) and may cause corrosion and mobilization of metals in metal pipes" (UNEP, 2002b).

8.2.5 Designing the gutter

In very low cost domestic rainwater harvesting, gutters can demand over half of the total investment (DTU, 2005c). Gutters in developing countries are frequently prone to failure due to poor design. They can contain leaks and can create a health hazard if water is allowed to remain in the gutter to attract mosquitoes. Gutter material used throughout the world includes plastic, aluminum, steel, wood, and bamboo. Each material has advantages and disadvantages. For example, wood and bamboo are usually cheap and all of the money towards its obtainment tends to stay within the community. However, these materials will eventually leak and degrade, and the porous surface attracts bacteria that may be washed into the rainwater storage tank (DTU, 2005c).

8.2.5.1 Water flow

The flow of water in a long gutter can be approximated by Manning's formula as follows:

$$Q = Av = A \times \frac{1}{n} R^{2/3} S^{1/2} \tag{8.6}$$

where

Q = flow in channel ($m^3\ s^{-1}$)
A = cross-sectional area (m^2)
V = velocity of flow in channel ($m\ s^{-1}$)
n = manning roughness coefficient
(usually between 0.01 and 0.015 for gutter material)
P = wetted perimeter (m)
R = hydraulic radius (m) [R = A/P]
S = slope of gutter

Using the Manning formula, the size of the gutter needed can be developed for any gutter profile (DTU, 2005c). Figure 8.5 shows the gutter material width (width of unfolded sheet material) needed for the most popular gutter profiles; the shaded area shows the range of typical domestic roof sizes.

Using the graph in Figure 8.5 the material width, gutter width at top, and cross sectional area of the gutter opening can be determined for various roof areas. The Table 8.8 gives gutter characteristics for the various buildings in

Laikipia using a semicircular (half round) gutter type since it will be the cheapest (least material use). The semicircular gutter shape is also used for the RWH at the Il Motiok Primary School (Figure 8.6). To illustrate how the graph is used to find the different parameters, a household roof of 60 m^2 will be used as an example. From the graph, one can find that a roof of 60 m^2 corresponds with a material width of approximately 150 mm. The gutter width at the top of a semicircular-shaped gutter is equal to the diameter of the semicircle. We can find this diameter as follows:

$$(0.5 \times 2)\ \pi \times r = 150 \text{ mm}$$
$$\pi \times r = 150$$
$$r = 150/\pi$$
$$r = 48 \text{ mm}$$

where r is radius

$$d = 2 \times r$$
$$d = 96\,\text{mm (gutter width at top)} \quad (8.7)$$

where d is diameter

We can find the cross-sectional area by using one half of the area of a circle.

$$\text{Cross sectional area} = 0.5 \times \pi \times r^2$$
$$= 0.5 \times \pi \times (46\,\text{mm})^2 \quad (8.8)$$
$$= 3600\,\text{mm}^2\ (36\,\text{cm}^2)$$

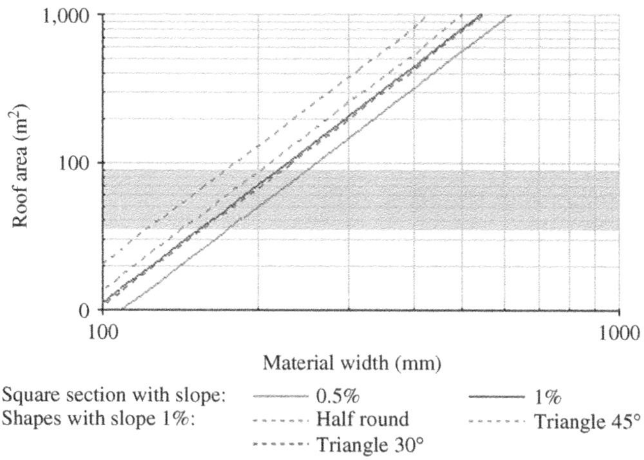

Figure 8.5 Material required for gutters to service various roof areas (DTU, 2005c)

Table 8.8 Gutter characteristics for buildings in Laikipia

Structure (Roof Areas rounded to nearest 10)	Material Use (mm)	Gutter Width (at top in mm)	Cross Sectional Area (cm^2)
Home in Il Motiok	115	73	21
Chief of Lekiji's House	125	80	25
Lekiji Primary School a	125	80	25
Lekiji Primary School b	160	102	41
Lekiji Primary School c	170	108	46
Lekiji Teacher Homes d	150	96	36

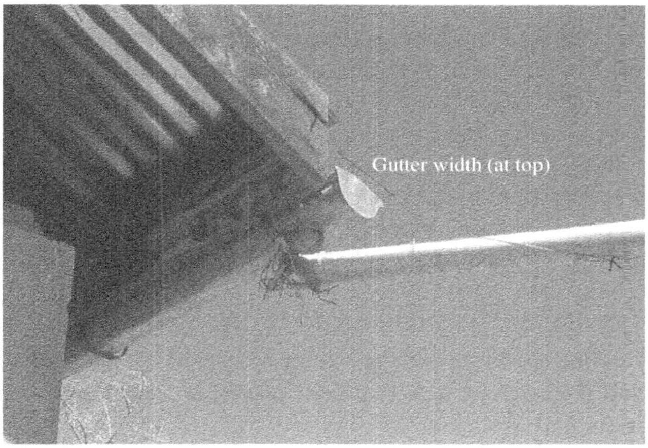

Figure 8.6 Gutter width at top from Il motiok primary school gutter

8.2.5.2 Interception

When water travels from a roof, it can take many paths depending on the intensity of rainfall and roof material. Winds can also affect the path of water coming from a roof. Studies from the University of Warwick indicate that for rainfall intensities between 0 and 2 mm/minute, an interception technology to direct the rainwater into the gutter is necessary (DTU, 2005c). They have identified three types of interception methods known as a splashguard, upstand, and enclosed gutter (Figure 8.7). A splashguard is a downward pointing sheet metal at the lip of the roof which directs the water vertically downwards into the gutter. An upstand raises the interception height of the gutter; an enclosed gutter completely surrounds the gutter-roof junction but makes the gutter difficult to clean and evaporation of any retained water less efficient.

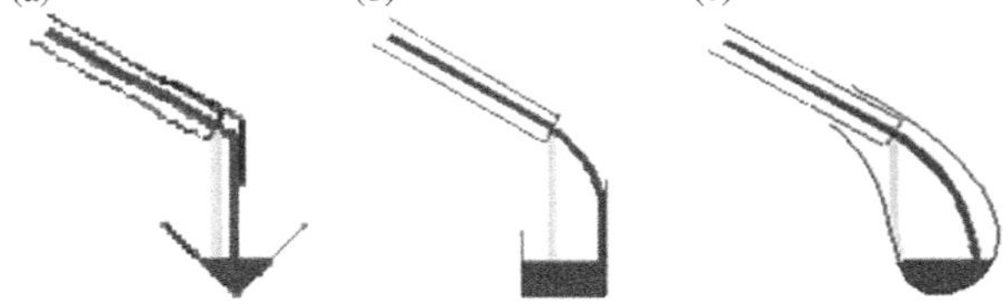

Figure 8.7 Roof interception methods (DTU, 2005c)

8.2.5.3 Mounting

Mounting the gutter to the roof can present problems; the mounting system must account for uneven roof edges, missing fascia boards, and other roof inconsistencies. The roof edge is a good place to mount the gutter since the gutter will automatically follow any lateral movement of the roof. This type of mounting is normally cheap because only wire is usually required. The disadvantage is that mounting the gutter so that it is firm and resistant to winds can be difficult. A common method for gutter mounting is connecting a fascia board to the roof edge; brackets can be mounted to the board in order to attach the gutter the board.

8.2.5.4 Pipes for gutter connection to storage tank

Table 8.9 from the Centre for Science and Environment in New Delhi gives an idea about the diameter of pipe required for draining out rainwater based on rainfall intensity and roof area:

Table 8.9 Diameter of rainwater pipe (in mm) for roof drainage (Centre for Science, 2005)

Roof	Average rate of rainfall in mm/h					
Area (m^2)	50	75	100	125	150	200
50	13.4	8.9	6.6	5.3	4.4	3.3
65	24.1	16.0	12.0	9.6	8.0	6.0
75	40.8	27.0	20.4	16.3	13.6	10.2
100	85.4	57.0	42.7	34.2	28.5	21.3
125	–	–	80.5	64.3	53.5	40.0
150	–	–	–	–	83.6	62.7

Note: mm/h – millimeters per hour; m – meters.

8.2.6 Technology to prevent and treat the contamination of rainwater

Figure 8.8 shows a schematic diagram that summarizes the contamination paths in RWH storage tanks and ways in which the contaminants can be reduced along the path. The discussion that follows describes the main low-cost technologies that prevent and treat contaminants in rainwater, including filters, first-flush devices, fine filtering, boiling, chlorination, and solar treatments.

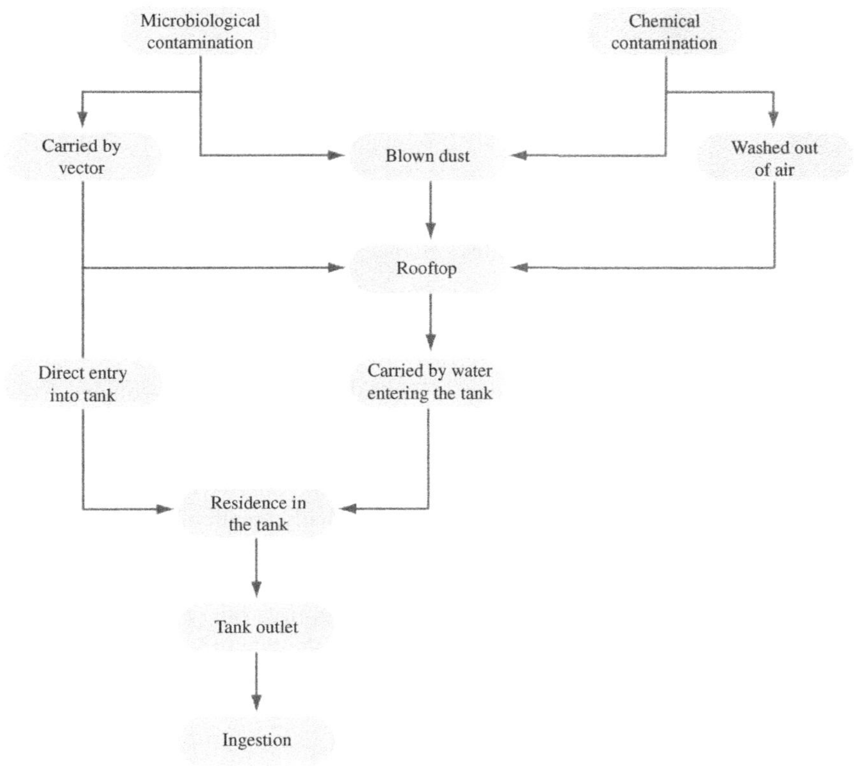

Figure 8.8 Contamination paths for roofwater harvesting (Thomas and Martinson, 2007)

8.2.6.1 *Filters*

It is necessary to install filters for the collection of rainwater in order to prevent contamination. Filters can be installed at the inlet to the storage tank, outlet, or both. To filter out leaves and large debris, a "leaf filter" can be installed. The most common positions for a leaf filter are at the gutter, at the beginning of the downpipe, in the downpipe, and at the entrance to the tank (DTU, 2005b). Figure 8.9 shows various types of filters used throughout the world, each with its own advantages and disadvantages.

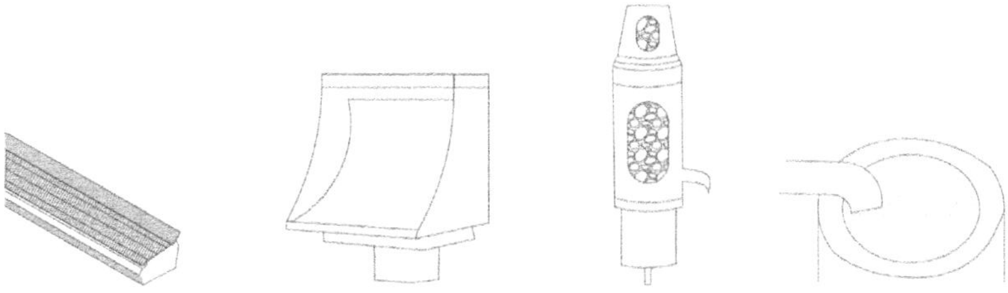

Figure 8.9 Various types of coarse leaf filters (DTU, 2005b)

8.2.6.2 *First-flush devices*

In order to divert the first flush, one can install a manual, fixed volume, fixed mass, or flow rate first-flush device. The manual device relies on someone being home and available to move the device in and out of place so that rainwater is directed away from the tank initially (Figure 8.10). The fixed volume method involves water filling a chamber of a set size (normally length of downpipe) until it overflows. The method can be used either with or without a floating ball seal which helps in reducing mixing between early dirty water and later clean water. The fixed mass system involves a mass of water to tip a bucket or seesaw. One type of flow rate system involves the balancing of the rate of water intake into a suspended hollow ball against its leakage, raising its weight and stretching its suspension until it descends into a recess, blocking the opening and allowing water into the tank (DTU, 2005b).

8.2.6.3 *Fine-filtering*

In order to filter out small sediments in the rainwater, one can install fine screens or sand filters. A filter unit is a chamber filled with filtering media (such as fiber, sand, charcoal, and gravel) in layers. Sand filters can effectively remove turbidity (suspended particles like silt and clay), color and microorganisms.

Charcoal can be used to provide carbon filtration. Figure 8.11 shows a simple filter made of gravel, sand, and charcoal and a simple sand filter. One type of sand filter was installed in Sri Lanka with success. The filter consisted of a 300 mm diameter tube filled with 150 mm sand on a bed of 200 mm of pebbles. It can deal with all except the very highest peak flows. It is important to clean the fine filters since they become dirty with particles trapped in them. If they are not cleaned, these particles will build up and eventually block the passing of water thereby creating overflow.

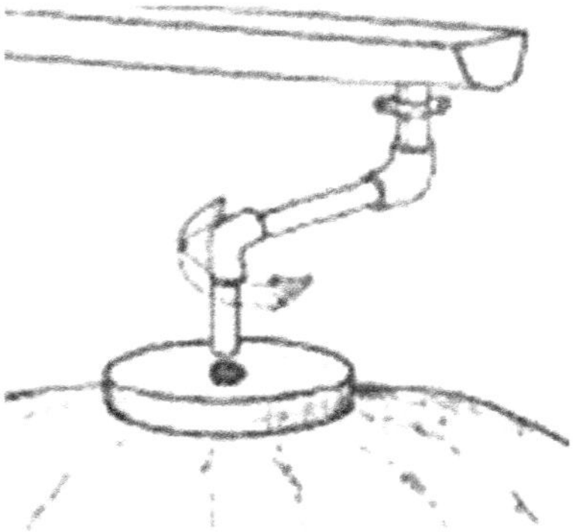

Figure 8.10 Manual first flush device (DTU, 2005b)

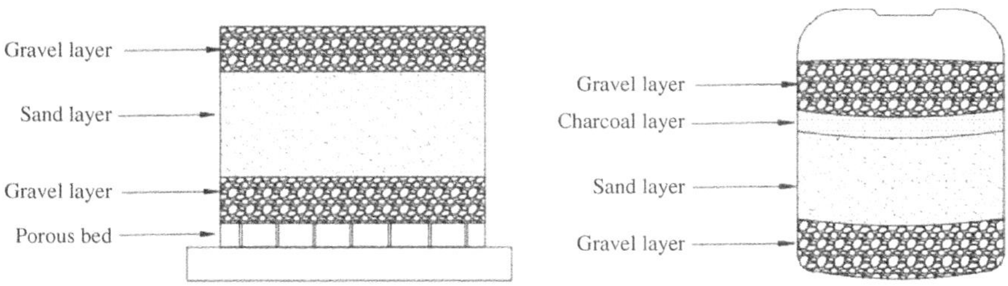

Figure 8.11 Gravel, sand, and charcoal filter (Centre for Science, 2005)

8.2.6.4 Bacteria in tank

A great deal of processes can occur when rainwater is stored in a tank, including sedimentation, floatation, and bacterial die-off. During sedimentation, the particulate matter in the water will settle to the bottom of the tank since it is heavier than the water. Lighter material will float at the top of the tank and is washed out with overflow water. Typical die-off behavior for microorganisms in water follows the pattern of a short period where numbers remain constant followed by exponential decline in numbers. The adverse environmental factors in the tank exceed the supportive factors due to the removal of organisms from their natural environment (DTU, 2005b). Sometimes there may be a short-term increase in bacterial numbers as the microorganisms take up residence. Residence time can be defined as the average time period during which a substance remains in its active form in a certain environment. The main factors for bacterial decline are: algae die off from lack of sunlight, increase in competition for food, increased predation reducing the prey microorganisms and ultimately starving out the predators, and flocculation and sedimentation causing the removal of some bacteria (DTU, 2005b). The level of die off can be calculated using the equation:

$$t = \frac{\ln(\frac{C_0}{C})}{k} \tag{8.9}$$

Where:

t is the elapsed time
C_0 is the initial bacterial concentration
C is the concentration at time t
k is a constant which depends on local factors such as UV levels, temperature and pH.

8.2.6.5 Boiling, chlorination, and solar treatment

Boiling water for 3–5 minutes kills pathogenic viruses, fungi, and bacteria, but this method is impractical on a large scale, and expensive (Macomber, 2001). In the field interviews in Kenya, many people stated that they did not boil water even if they perceived it to be unclean.

Chlorination is widely recommended as a final sterilization for rainwater, but it is not generally well liked by users and the chemicals used can be dangerous if misused (DTU, 2005b). In *Guidelines on Rainwater Catchment Systems for Hawaii*, chlorine treatment for rainwater storage tanks is outlined. One should add 2.5 ounces (about 71 grams) of unscented laundry bleach (containing 5.25 percent available sodium hypochlorite) for every 1,000 gallons (3,785 L) of water during rainy periods, and add 5 ounces (about 142 grams) per 1,000 gallons of water during dry periods (Macomber, 2001). If there are biological pollutants in the system, they tend to become more concentrated in dry periods, and more chlorine is needed to disinfect the water. The idea is to maintain 1 part per million (1 ppm) of chlorine in the tank. In the absence of a method of precisely monitoring and maintaining this level of chlorine in the tank, some sources recommend that chlorine be added once a month.

Pasteurization is another low-cost method of providing clean drinking water. The pasteurization process heats the water to a temperature high enough to kill bacteria, viruses, and other water-borne pathogens. There are several methods for pasteurizing water using solar radiation, including using solar cookers. Table 8.10 presents the advantages and limitations to solar water pasteurization.

Table 8.10 Advantages and limitations to solar water pasteurization (Safe Water Systems, 2005)

Advantages	Limitations
99% effective in destroying water-borne pathogens	Requires sunny weather
Uses only the heat of the sun; does not use electricity, chemicals or fossil fuels	Cannot operate in freezing conditions
Prevents air pollution, ozone damage, deforestation, and soil erosion	Does not remove chemical pollutants from water
Alternative to chlorine treatment, which can be ineffective at destroying worms & protozoa	Does not remove minerals or salt from water (does not desalinate water)
Low maintenance	

Solar Water Disinfection (SODIS) is a simple technology used to improve the microbiological quality of drinking water. SODIS uses solar radiation to destroy pathogenic microorganisms by filling contaminated water into transparent plastic bottles and exposing them to full sunlight for six hours. The most favorable region for SODIS lies between latitudes 15°N/S and 35°N/S. These semi-arid regions are characterized by high solar radiation and limited cloud coverage and rainfall (3,000 hours sunshine per year) (EAWAG, 2005). Limitations of SODIS include the fact that it does not change the chemical water quality, requires relatively clear water (turbidity less than 30 NTU), requires suitable weather conditions, and is not useful to treat large volumes of water (EAWAG, 2005). According to EAWAG, the following instructions are given for solar water disinfection:

(1) Wash the bottle well the first time you use it.
(2) Fill the bottle fully and close the lid.
(3) Place the bottles on a corrugated iron sheet or put them on the roof.

(4) Expose the bottle to the sun from the morning until evening for at least six hours.
(5) The water is now ready for consumption.

8.2.7 Efficiency

It is important to analyze the performance of the RWH system once it is in place. Although this project only deals with preliminary design considerations for a RWH system for the buildings near MRC, a brief discussion of efficiency is provided here.

The efficiency of the RWH system can be analyzed once it is in place. An inefficient storage volume of rainwater can be due to poor guttering, leaking tanks, and an unsuitable storage tank size. An inefficient tank can be defined as one in which a significant fraction of rainwater is lost to overflow. Unreliability can be expressed as the fraction of time when the storage tank is dry or the fraction of annual water use that has to be taken from other water sources (DTU, 2005e).

Given that water inflow into a storage tank is less than the water demand, the storage efficiency can be calculated as:

$$\textit{Storage efficiency}\ (\%) = 100 \times (1 - \textit{overflow/inflow})$$

The system efficiency can be calculated as:

$$\textit{System efficiency}\ (\%) = 100 \times \textit{water used/water falling on the roof}$$

8.3 SUMMARY

This chapter provided information relevant to designing domestic RWH systems. Most importantly, this chapter presented designs for six buildings near the MRC in Laikipia, Kenya. Because of the drought conditions in this region, RWH can most reasonably be designed as a supplement to other water resources. The corrugated galvanized iron roofs, which are currently present on many of the buildings in Laikipia, provide a good runoff coefficient of 0.9 to facilitate rainwater collection.

The supply of rainwater was discussed using yearly, monthly, and daily rainfall data. Using rainfall data averaged over a year, and the United Nations definition of the average minimum amount of daily water needs per person (50 L per person per day), it was found that the six buildings could generally provide enough water to meet *drinking* water needs (5 L per person per day). Different designs for the rainwater storage tanks were also determined using an average of two sets of 15-day consecutive rainfall events that occurred in November of 2001 and in April of 2003. The maximum storage tank volume was found using the following: the daily rainfall data, the area of the horizontal projection of the various roofs, the runoff coefficient for the roof, and a pre-determined daily water demand/consumption that was based on water needs. The maximum storage volume was set to be the design size for the tanks; the storage tank sizes range from 1,900 L to 10,000 L for the six buildings. The amount of time it takes for the storage volume of water in each tank to empty was also explored; a general average of a month of water could be supplied after the initial 15-day rainfall event occurred.

The gutter sizes for the six buildings were also determined using an available plot of roof area versus material width. Finally, important considerations of the RWH system, such as technologies that prevent and treat contaminants in rainwater (like filters, first-flush devices, fine filtering, chlorination, and solar treatments) were discussed. Filters and first-flush devices can be effective at reducing contamination; first-flush mechanisms can be manual or automated. While chlorination and solar water treatment can be effective at destroying bacteria, chlorination can be dangerous if misused, and solar water treatment is restricted to certain locations with adequate sunlight.

9
Policy considerations and conclusions

9.1 INTRODUCTION: POLICY CONSIDERATIONS AND CONCLUSIONS

This chapter provides summaries, conclusions, and policy recommendations based on the findings presented in chapters one through eight. Specifically, the chapter describes the similarities and differences between the Panama and Kenya case studies, which include region-specific considerations, issues relevant to dry climates, and issues relevant to wet climates. In addition, a brief discussion of feasible low cost technologies is provided. Policies that encourage sustainable building water practices are explored; environmental regulations from Panama, Kenya, and other countries with laws that promote sustainable water resources are presented. Finally, policy recommendations

ISBN: 9781843393238. Published by IWA Publishing, London, UK.

regarding the sustainable water resources in the built environment in Panama, Kenya, and developing countries in general are provided.

9.2 SIMILARITIES AND DIFFERENCES BETWEEN CASE STUDIES

Similarities and differences can be found in the Panama and Kenya case studies with respect to wastewater, potable water, budget, climate, culture, sustainable building usage, and water regulations.

9.2.1 Wastewater

The two case studies deal with two different ways of treating wastewater or human waste from a building, and show inadequate waste/wastewater systems in the surrounding areas. In Panama, septic tank systems are currently in use at STRI. However, there are plans to construct a wastewater stabilization pond and constructed wetland system at STRI to treat the wastewater, including excreta from flush-toilets. The use of flush-toilets (as opposed to composting ones) is more appropriate in Panama than Kenya due to Panama's abundant rainfall and overall adequate water quantity. Based on the analysis of water issues in the Kenya case study, it is recommended that MRC and its surrounding communities incorporate composting eco-toilets at the household and school level. The use of composting toilets for MRC will save water in the rainfall-deprived area and help to decrease the level of pollution in nearby surface waters. Although the surrounding villages of MRC use pit latrines which do not use water, composting toilets such as the Fossa Alterna type can (1) provide useful fertilizer to grow crops and (2) provide less contaminated land sites that arise each time a new pit is dug for a pit latrine.

In both countries, there were some cultural barriers to the handling of human excreta with the use of composting toilets were found. However, this barrier seemed to be stronger in Panama than in Kenya. In Kenya, the chief of Lekiji and many influential members of the communities that were visited were in favor of composting toilets. In both case study areas, the surrounding communities had generally poor sanitation facilities; in Panama, much of the waste from the latrines went straight into the ocean and the local wastewater treatment pond was leaking contaminants into the soil, while in Kenya, many of the pit latrines had strong odors among other problems.

9.2.2 Potable water

In regards to potable water, the two case studies reveal the potentials and constraints of RWH, as well as demonstrate examples of local water resource problems. In Panama, the municipal potable water treatment plant was generally inadequate in providing consistent good-quality water. In Kenya, many people in the villages used the contaminated Ewaso Nyiro or Nanyuki River water for potable purposes. In both locations, rainwater stored in storage tanks was found to be contaminated with total and fecal coliforms. Thus, independent of the location, rainwater collected from roofs has the risk of being contaminated from wildlife and subsequently needs to incorporate prevention and/or treatment methods. In neither of the case studies did the rainwater storage tanks tested contain rainwater from a collection system that used a first-flush device. The use of this type of mechanism, while inconsequential for the high-budget science building's rainwater treatment system in Panama, could be very vital for low-budget RWH systems at the household level.

9.3 REGION-SPECIFIC CONSIDERATIONS

The analysis of the water aspects of sustainable construction in Panama and Kenya illustrate the need to be region-specific (and even site specific) when designing and implementing sustainable building water solutions. One must be alert to *all* environmental factors when constructing green buildings in developing countries; this includes the social, political, and physical environment.

9.3.1 Issues for dry climates

Dry climates such as Kenya often suffer severe water shortages and scarce water resources. This often leads to conflict over water. In *Kenya at the Cross-roads*, the conflict in the Ewaso Nyiro River basin (east side of MRC) that arises from sharing water between irrigation and pastoralist activities is described. Expansion of horticultural production upstream denies downstream pastoralists sufficient water, and the Lorian Swamp cannot be replenished adequately (which threatens livestock survival) (IEA and SID, 2001). Figure 9.1 shows various violent outbreaks that took place in Kenya in 2005. In January 2005 in Mai Mahiu, Nakuru, there were 2,000 displaced families and 14 people killed following violent clashes between two communities over water. Figures 9.2

and 9.3 show information regarding the Kenyan droughts in 2004 and 2006. All of these figures point to the extreme importance of water supply to the stability of communities. Sustainable buildings, through the use of water collecting and saving mechanisms, may help to alleviate this problem. Thus, RWH and composting toilets are two critical components of buildings in dry climates.

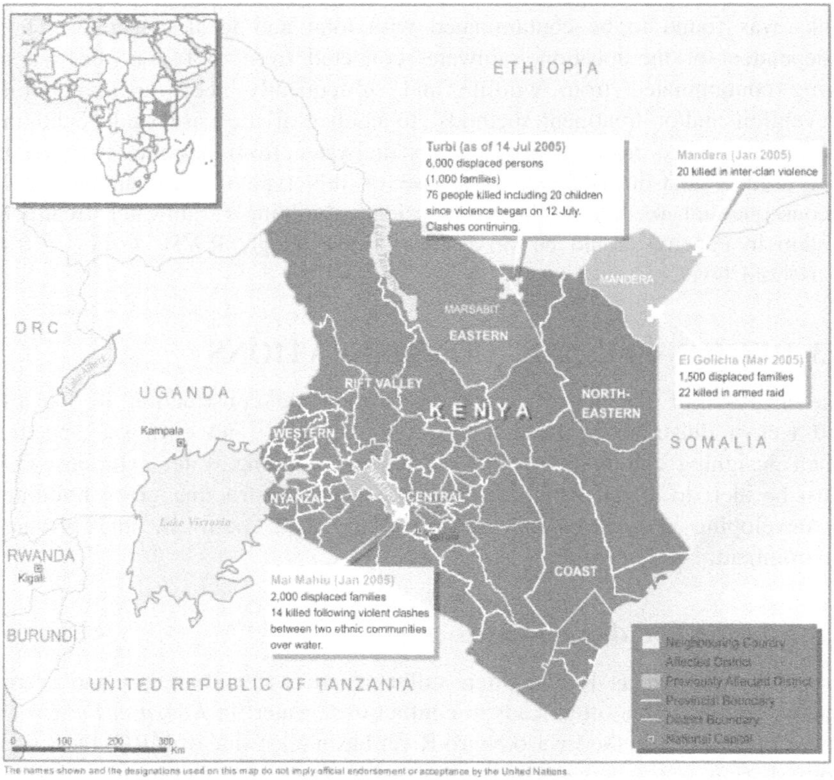

Figure 9.1 July 2005 Violent outbreaks in Kenya (ReliefWeb, 2005). Map provided courtesy of the ReliefWeb Map Centre, UN office for the coordination of humanitarian affairs. The boundaries and names shown and the designations used on this map do not imply official endorsement or acceptance by the United Nations

The maps in Figures 9.2 and 9.3 show (1) problems in Kenya brought by the 2004 drought and (2) the drought situation as of January 2006.

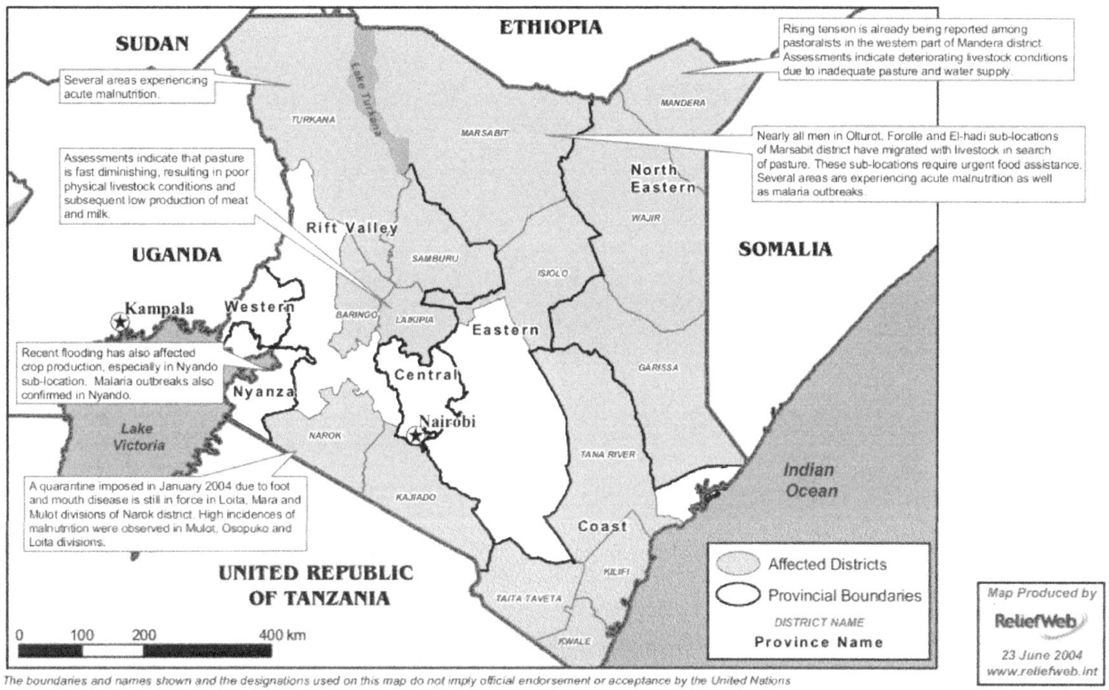

Figure 9.2 Kenya drought 2004. (ReliefWeb, 2004). Map provided courtesy of the ReliefWeb Map Centre, UN office for the coordination of humanitarian affairs. The boundaries and names shown and the designations used on this map do not imply official endorsement or acceptance by the United Nations

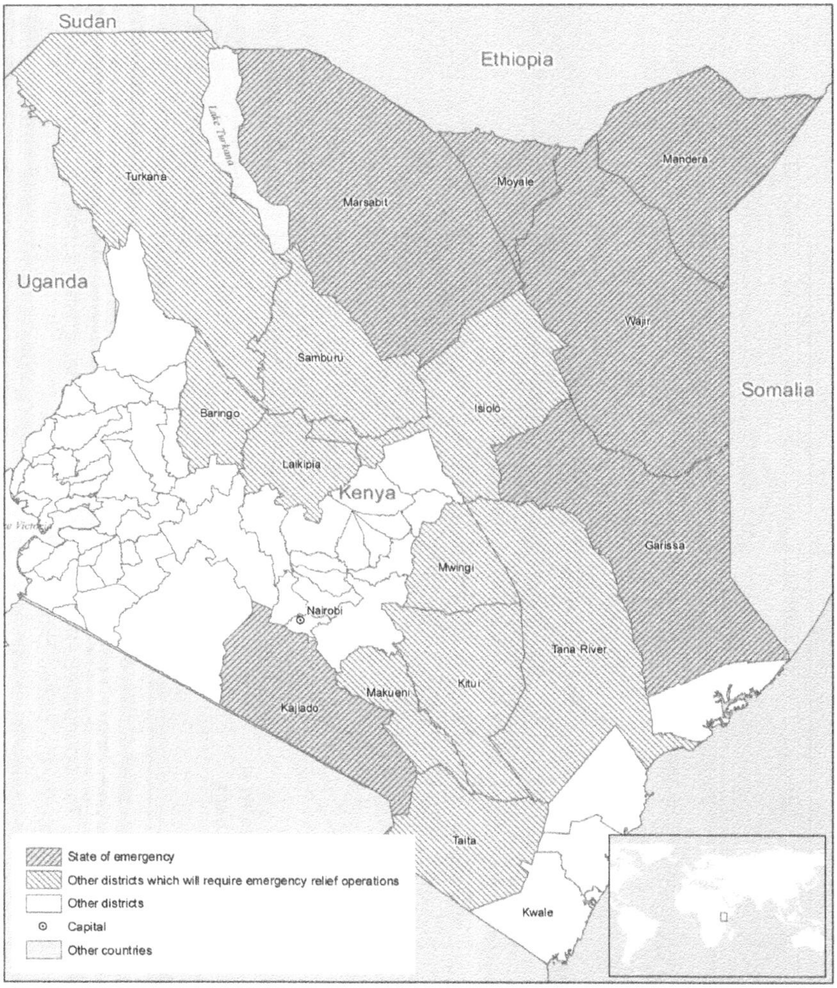

Figure 9.3 Kenya drought 2006 (IFRC, 2006). The maps used do not imply the expression of any opinion on the part of the International Federation of Red Cross and Red Crescent Societies or National Societies concerning the legal status of a territory or of its authorities

9.3.2 Issues for wet climates

Wet climates enjoy plentiful rainfall, where rainwater harvesting from buildings can provide water for full usage. Rainwater harvesting can be especially beneficial in marginalized areas that are not linked to adequate water systems, such as Bocas del Toro Panama. Although the need to conserve water is less critical in wet climates, the use of composting toilets still serves a good purpose as it turns waste into an economic and food security benefit. Too much rain can be a problem from an economic and design perspective. One designs the size of a rainwater storage tank using the roof size and rainfall data for the region. However, as the size of the tank increases, so does the cost of it. It is possible that one may not be able to afford a tank big enough to hold the water, thus limiting the full usage of the rainwater and creating a lot of overflow if a small tank is used.

9.4 FEASIBLE LOW COST TECHNOLOGIES

In order to construct sustainable buildings in developing countries, the use of low-cost technologies must be employed. The materials that are least expensive will depend on the region. Material costs, shipping costs, labor costs, and maintenance costs, should all be considered, and costs should be weighed against the benefits of the technology. For example, although RWH tanks can be costly, they may well be worth it, especially in drought-prone areas such as Laikipia, Kenya. Although individual RWH systems are more favorable than public communal ones, community systems, such as RWH for local schools, may be the only feasible solution for some due to financial constraints.

9.5 POLICIES THAT ENCOURAGE SUSTAINABLE BUILDING WATER PRACTICES

Developing countries need environmental policies that support the concept of sustainable development through sustainable construction. There are currently many examples of country policies that encourage sustainable water practices, which include tax incentives, implementation mandates, and other creative methods.

In India, legislation on rainwater harvesting includes mandates and tax incentives. Various regions of India have mandated that all new buildings with roof areas of a certain size must have RWH. In New Delhi, for example, all new building with a roof area of more than 100 square meters must include RWH;

in Haryana, RWH is mandatory for all new buildings irrespective of roof area (Centre for Science and Environment, 2005b). In Indore, a rebate of six percent on property tax is available for implementing RWH systems.

Unfortunately, in neither Kenya nor Panama, are there specific regulations that provide mandates or incentives for implementing sustainable water practices (such as RWH, on-site wastewater recycling, and composting toilet systems) into buildings. The description below highlights some of the legislation relating to water in both countries.

9.5.1 Panama

When the United States was building the Panama Canal from 1904 to 1914, Panama followed U.S. standards for water quality. In the year 2000, Panama created new Panamanian environmental regulations with the Pan-American Health Organization (PAHO) playing a major role. The three main categories for the laws set in 2000 are watersheds, drinking water, and wastewater. Laws concerning the water standards were set to be in accordance with the World Bank and Inter-American Development Bank (IDB) in order for development funds to be awarded in the future from those organizations.

According to Civil and Environmental Engineering Professor Casilda Saavedra at the Technological University of Panama, Panama's biggest *environmental* problem is water pollution in rivers from both solid waste and wastewater (Saavedra, 2005). Fortunately, the major potable water supply sources are not polluted (such as Madden Lake, Gatún Lake, Las Cabras River), and most of Panama boasts adequate drinking water quality. However, it appears that marginalized areas in the far west and far eastern provinces (Bocas del Toro and Darién, respectively) do not enjoy a clean water supply. These areas therefore can benefit from proper rainwater harvesting. Many low-cost rainwater harvesting systems have been in place for hundreds of years in these areas.

The Smithsonian Institution's regulations for their overseas stations stipulate that the water quality for the station must follow whichever law is stricter, whether it is the host country or the USEPA policies. The Panamanian regulations for potable water are very similar to those of the USEPA and WHO. Table 9.1a presents a comparison among USEPA, WHO, EC, and the Panamanian government of standards for those water quality parameters measured in the Panama experiment. The Panamanian standards in Table 9.1a for fecal and total coliform pertain to *rainwater* potable water. Panama sets different standards for the various types of potable water as Table 9.1b indicates.

Table 9.1a Panama, WHO, USEPA, and EC potable water standards

Parameter	Unit*	WHO Guideline	USEPA Max Contamination Level**	Panama Regulations	EC Max Admissible Concentration***
pH	pH units	6.5–8.5	6.5–8.5 ◇	6.5–8.5	6.5–8.5
TSS	mg/L	NS	NS	NS	NS
Turbidity	NTU	5	1 ○	1	4 FTU
Fecal Coliform	**CFU/100 mL**	**0**	**<1** □	**0**	**0**
Total Coliform	**CFU/100 mL**	**0**	**<1** □	**10**	**0**
Sulfates	mg/L	250	250 ◇	250	250
Nitrates	mg/L	50	10	10	50
Nitrites	mg/L	3	1	1	0.1
Color	Cu	15	NS	15	20 mg Pt-Co/L

NS No Standard
* *Unit Unless Otherwise Noted*
** United States Environmental Protection Agency
*** *European Economic Community (established by European Committee for Environmental Legislation)*
◇ United States National Secondary Drinking Water Regulations
○ The turbidity limit is 1 NTU at any given time; the turbidity is not to exceed 0.3 NTU in 95% of water samples for a given month
□ No more than 5% of water samples can test positive for the presence of total coliform in a given month. If total coliforms are present, fecal coliforms are tested to determine presence/absence. Absence is declared if there is less than 1 CFU/100 mL; presence is declared if they are equal to or greater than 1 CFU/100 mL

There are regulations in Panama concerning the use of compost for commercial and domestic fertilizer. These regulations encompass all of the establishments treating wastewaters and every kind of wastewater treatment plant that generates sludge as a result of the treatment process. The regulations are applied to individuals or businesses that (1) are involved in the handling of sludge and their commercialization, whether directly or as a subproduct, (2) apply sludge to agricultural fields, and (3) are dedicated to the cleaning and extraction of material, whether liquid or sludge, from septic tanks or pits, residential or industrial. The regulations thus fail to describe the use of household composting toilets, as well as regulations and market for that compost.

Table 9.1b Panamanian biological standards for potable water (ministry of commerce and industry of Panama, 1999)

CHARACTERISTICS	UNIT	MVP*	OBSERVATIONS
1. Biological			
A. Water distributed by pipes			
A.1 Treated water that enters the distribution system			
Fecal coliform bacteria	N of colonies/100 mL	0	
Total coliform bacteria	N of colonies/100 mL	0	
A.2 Untreated water that enters the distribution system			
Fecal coliform bacteria	N of colonies/100 mL	0	
Total coliform bacteria	N of colonies/100 mL	3	
A.3 Water in the distribution system			
Fecal coliform bacteria	N of colonies/100 mL	0	
Total coliform bacteria	N of colonies/100 mL	3	
B. Water not distributed by pipes			
Fecal coliform bacteria	N of colonies/100 mL	0	
Total coliform bacteria	N of colonies/100 mL	10	
C. Bottled water			
Fecal coliform bacteria	N of colonies/100 mL	0	The source of the water supply should be as free from fecal contamination as the water at the end of the process.
Total coliform bacteria	N of colonies/100 mL	0	
D. Parameters of fulfillment for all conditions (A, B & C)			
Protozoans (pathogenic)			The actual knowledge gathered is not such that guideline values can be established for biological characteristics. However, the presence of any of these organisms in potable water is indicative of contamination and is a cause of sickness. Therefore, such organisms should not be present in water meant for human consumption.
Helminths (pathogenic)			
Free-living organisms (algae, seaweed, other)			
Enterovirus			
Other organisms			

*Maximum Value Permitted

The following rules presented are from Panama's Technical Rules DGNTI-COPANIT 47–2000 (Ministry of Commerce and Industry of Panama, 2000). The use of *industrial* sludge for agricultural use, production of fertilizer for agricultural, urban, or recreational use, is absolutely prohibited. For the transformation of sludge into fertilizer, sludge can be mixed with other residue and/or materials, such as organic residue from gardens. This material can be used in the following cases: fertilizer for horticulture, additive for decorative plant nurseries, additive for physical improvement of soils, and fertilizer for recreation areas, such as parks and golf courses. If sludge from wastewater treatment processes is to be used as a fertilizer or for agricultural applications, the treatment process must include one or more of the processes in **Classes I and II**. Sludge must also comply with the maximum allowable limits of heavy metals specified in Tables 9.2 and 9.3. Treatment Methods of Class I **(Sludge Class I)** includes the following treatment processes: aerobic digestion, anaerobic digestion, air drying, transformation of sludge into fertilizer (composting), stabilization, and other treatment processes. Treatment Methods of Class II **(Sludge Class II)** includes the following treatment processes: transformation into fertilizer as in Class I, sludge drying with heat, thermophilic anaerobic digestion, pasteurization, and other treatment processes.

Table 9.2 Maximum allowable limits for heavy metals and fecal coliforms for sludge to be used as a fertilizer (Ministry of Commerce and Industry of Panama, 2000)

Parameters	Maximum Allowable Limit	Units (based on dry weight)
Arsenic	75	mg/kg
Cadmium	85	mg/kg
Chromium	3000	mg/kg
Copper	4300	mg/kg
Lead	840	mg/kg
Molibdenum	75	mg/kg
Mercury	57	mg/kg
Nickel	420	mg/kg
Selenium	100	mg/kg
Zinc	5	mg/L
Fecal coliforms	2000	CFU/gram
pH	9–12	

Unfortunately, there is no mention of RWH or composting toilets in any of Panama's water management and water quality regulations.

Table 9.3 Maximum allowable limits for heavy metals and fecal coliforms for sludge to be used in agricultural applications (Ministry of commerce and industry of Panama, 2000)

Parameters	Maximum Allowable Limit	Units (based on dry weight)
Arsenic	40	mg/kg
Cadmium	40	mg/kg
Chromium	1500	mg/kg
Copper	1500	mg/kg
Lead	300	mg/kg
Molibdenum	25	mg/kg
Mercury	25	mg/kg
Nickel	420	mg/kg
Selenium	50	mg/kg
Zinc	3000	mg/L
Fecal coliforms	2000	CFU/gram
pH	9–12	

Table 9.4 shows the sampling frequency for sludge commercialization in Panama.

Table 9.4 Sampling frequency for sludge commercialization (Ministry of commerce and industry of Panama, 2000)

Amount of Sludge to be commercialized (tons/yr)	Minimum Control Frequency	Parameters
0 to 300	Once a year	Those specified in Tables 9.2 and 9.3
301 to 1500	Every three months	
1501 to 15000	Every sixty days	
>15000	Once a month	

9.5.2 Kenya

Kenyan water regulations mention RWH, but present unclear information regarding its use. The Kenyan "water bill" of 2002 clearly mentions that the country's water resources are vested with the state and the Minister in charge has the right to control every water resource. According to the Kenyan water bill, a permit is required to extract water from a water resource (except for domestic purposes) without "employment of works"; it is not clear whether this provision applies in the case of community RWH systems (De S Ariyabandu, 2003).

The water bill states that the Kenyan authority will not approve any community project unless the person(s) owning or occupying at least two thirds of the land approves the project. *These laws if interpreted to include RWH for households and communities can be a constraint to the development of the concept of rainwater harvesting for domestic use* (De S Ariyabandu, 2003).

Contrary to the water bill, the "National Policy on Water Resources Management and Development" explicitly mentions the development of domestic RWH in urban and rural areas as a source for domestic water supply. As was stated in *Very Low-Cost Domestic Roofwater Harvesting in the Humid Tropics: Its Role in Water Policy* by Rajindra De S Ariyabandu, "The [Kenyan] policy also gives recognition to community management of water resources and emphasizes the importance of decision making at various management levels including river basins and sub basins." Thus, two major water regulations seem to point to different directions regarding RWH and water management. The "policy" is clearly in support of domestic RWH as well as decentralized water management, while the "bill" is in favor of centralized control over water resources and is unclear on how RWH fits within that scope. In paragraph 251 of the Building code (city council by-laws) in Kenya, it is specified that action must be taken with respect to roof run off. The Building code specifies effective disposal of roof water but do not give any reference to storage and utilization of rainwater.

9.5.2.1 Ecotourism

Ecotourism can be defined as tourism in which the goal is to experience nature for nature's sake (Nagel, 2002). Advocates of ecotourism argue that this type of tourism is one of sustainable development since it is capitalizing on areas of natural history and environmental beauty. This approach seems to be providing great benefits in Latin America and Africa. For instance, Ecuador's Galapagos Islands' tourism grew by more than 335 percent between 1974 and 1987 (Nagel, 2002). Ecotourism has two main benefits, which are its contribution to economic development and a reduction in the pace of environmental destruction. On the other hand, negative outcomes can arise if the needs of residents living near an ecotourism location are compromised. Tourism is Kenya's primary foreign currency earner (Nagel, 2002). In Kenya, residents living near Amboseli National Park, which boasts massive tourism each year, felt that they were ignored by the Kenyan government. To remedy this situation, the government now provides fiscal incentives to communities by ensuring them that local goods, services, and labor are used by tourism whenever possible and a portion of the money from game lodges is allocated to the local communities.

In the Kenya case study, another issue surfaced with respect to ecotourism. The eco-tourism lodges in Il Motiok used thatch roofs to aesthetically please potential tourist. However, thatch is not appropriate for collecting rainwater and thus contaminated river water would have to be used for non-potable water purposes (while bottled water would be used for potable purposes). Although a total of 100 sq. m of roof catchment surfaces will be used to collect rainwater, this only allows for a water supply of 36,000–54,000 L in a year, or roughly 100–150 L per day (based on a runoff coefficient of 0.9 and mean annual rainfall ranging between 400 mm and 600 mm). Much more rainfall could be harvested if the lodges used a more appropriate roof material for RWH. Tradeoffs such as this—aesthetically pleasing tourist versus providing needed water—should be taken into consideration when constructing sustainable buildings for ecotourism in developing countries.

9.5.3 Policy recommendations

Water resources are a part of a larger system that includes health, habitat, food, the built environment, energy and climate. Any effort to isolate the problems of adequate water quantity and quality from the widely-defined ecosystem would lead to misguided conclusions. It was seen through this research, especially from the Kenya case study, that environmental management is intricately linked with human development. As was stated by Musalia Mudavadi in *Rethinking Development in Africa in the 21st Century: The Case of Kenya*, "The over-arching aim of development is to eventually eliminate poverty and improve the quality of life. This depends on environmental resources. Individual countries will not vigorously conserve the environment unless they have something to gain from it" (Mudavadi, 2002). Sustainable construction can be one of the means in which development aids the improvement of life while also helping countries to improve economic, social, and environmental conditions.

Mudavadi also discusses how sustainable development rests on political will, and that this "will" should be expressed in specific national policies on environmental sustainability. The following bullet points present policy recommendations on specific actions that can be taken (and are being taken in many parts of the world) to promote the water aspects of sustainable buildings. These are conclusions that arose from the analysis of the Kenya and Panama case studies.

- Dry climate regions such as Kenya should have water policies that ***provide incentives for the use of composting toilets***. Without using water to flush toilets, these drought-prone regions can save considerably on water quantity while also creating good-quality fertilizer to use on-site or sell.

- The market for fertilizer made from composting human waste should be explored and regulated in those regions such as Laikipia, Kenya that have little use for the compost (due to abundant cattle). The market for composted fertilizer from buildings (in addition to compost from wastewater treatment plants) should also be included in Panamanian regulations. Thus, Panama's Technical Rules DGNTI-COPANIT 47–2000 should be revised to encompass composting toilets.
- Dry climate regions such as Kenya and wet climate regions with poor water systems should have a combination of ***RWH mandates and tax incentives*** for new commercial and residential buildings. It should be noted that RWH may not be a panacea for water quantity and quality problems, but can help to remedy the issue especially as a climate change adaptation measure. As the University of Warwick School of Engineering: Development Technology Unit states, *There is a complex set of interrelated circumstances which have to be considered when choosing the appropriate water source. Cost, climate, technology, hydrology, social and political elements all play a role in the eventual choice of water supply scheme which is adopted for a given situation. [RWH] is a technology which is flexible and adaptable to a very wide variety of conditions, being used in the richest and the poorest societies on our planet, and in the wettest and the driest regions of the world* (University of Warwick, 2007).
- Wet climate regions with adequate water systems such as Panama should encourage the use of RWH, especially in those rural regions without adequate water distribution and treatment systems.
- RWH can be made more effective by guttering both long sides of a rectangular shaped roof, rather than guttering only one side as was seen in most buildings in the Kenya case study. Instead of guttering two sides with two separate rainwater storage tanks, it may be more cost-effective to have one large tank with gutters and downpipes from the entire roof connected to the tank.
- RWH can be made more effective by reducing the potential contamination with a first-flush device. This device can be an automatic or manually operated mechanism by which the pipe that connects the gutters to the storage tank (downpipe) is moved to the tank only after the first few minutes of a downpour. This allows the roof, which may contain biological and chemical contaminants, to be cleaned at least once before the rain begins to runoff into the storage tank.

- The chief of Lekiji's suggestion to use rainwater for potable purposes and use river water for irrigation purposes provides valuable insight into how to ration water resources appropriately in a drought-prone area. This rationing must be done according to local conditions.
- Information on the range of technologies, method of design and construction, and costs of RWH systems should be readily available to the public. This information should be specific to the region in question. For example, the prices of the various components of the RWH systems, which include roofing materials, gutters, conveyance pipes, first-flush diverters, and storage tanks, should be listed in country-specific terms. For instance, the price of concrete will be relatively inexpensive in some regions, while it is prohibitively too costly in other areas. This can be achieved through national campaigns in conjunction with international organizations, local non-governmental organizations (NGOs), and private voluntary organizations (PVOs). Target audiences for this information include schools, community organizations, and local leaders. The Kenya Rainwater Association (KRA), for example, has been active in lobbying for policies which support RWH.
- The Kenya and Panama case studies were both conducted through research centers, which are at least partly funded through private donations and developed countries. One way that these types of institutions can be of benefit to the surrounding communities, while they benefit from studying their habitat, is by helping to fund sustainable building projects. This may include providing corrugated iron roofs appropriate for rainwater harvesting to village homes in Kenya and providing cleaner water and sanitation facilities for community entities such as schools near STRI in Bocas del Toro, Panama.
- Sustainable buildings should be an integral part of national environmental policies. As Mudavadi stated, "Issues such as patterns of rainfall need to be factored into strategies designed to address citizens' social and economic needs" (Mudavadi, 2002).
- Sustainable buildings, used as a part of ecotourism in developing countries, should consider the needs of the local community in addition to what is aesthetically pleasing to tourists. Tourists can benefit and may be happy to contribute to the local environment, economy, and community if they are given information regarding how their buildings and activities positively benefit the community.
- Authors Falkenmark and Rockström have listed three efforts necessary for successful water management: (1) secure water and related services, (2) avoid degradation of water, land resources and ecosystem integrity,

and (3) foresee changes (climate, population, diet preferences, etc.) (Falkenmark, 2007). By incorporating water supply and savings mechanisms in the built environment, one can help to address all three of these requirements.

- Water demand is expected to increase over time due primarily to population increase and increases in average standards of living. At the same time, climate change exacerbates water problems in many parts of the world. According to the IPCC, adaptation options designed to ensure water supply during average and drought conditions require integrated demand-side as well as supply-side strategies (Bates, Kundzewicz, Wu, & Palutikof, 2008). This recommendation is validated in the Laikipia and Bocas del Toro case studies (supplying water through RWH while decreasing water demand through composting toilets, for example).
- Sustainable water resource management is increasingly important as climate change adaptation and mitigation measures gain momentum. The work contained in this publication thus serves as part of a response to adapt to a changing climate, with case studies relevant to both wet climates and dry climates. It will be important for policy makers to consider how the methods of RWH and ecological sanitation help to achieve development objectives in a greenhouse-constrained world.

10
References

Airy, M.W. (2005). *Report on Water Supplies for Mpala Research Centre.* Laikipia: Mpala Research Centre.

Albrechtsen, H. (2002). Microbiological investigations of rainwater and graywater collected for toilet flushing. *Water Science and Technology*, **46**(6–7), 311–316.

Allen, D. (2005, November 4). (M. Waite, Interviewer) personal interview on date stated.

Andersson, I. (2005). *Sewage Becomes the Solution not the Problem.* Retrieved August 2005, from UNDP: http://www.undp.org/dpa/choices/18june99/index.html.

Australian Government. (2004). Retrieved January 2, 2006, from Guidance on use of rainwater tanks: http://enhealth.nphp.gov.au/council/pubs/documents/rainwater_tanks.pdf.

Baker, G.A. (2005, November 1). (M. Waite, Interviewer) personal interview on date stated.

Bale, J. and Drakakis-Smith, D. (Eds). (1988). *Ecology and Development in the Third World.* New York: Routledge.

Bates, B., Kundzewicz, Z., Wu, S., and Palutikof, J. (2008). *Climate Change and Water: Technical Paper of the Intergovernmental Panel on Climate Change.* Geneva: IPCC Secretariat.

Battle, G. and McCarthy, C. (2001). *Sustainable Ecosystems and the Built Environment.* New York: John Wiley and Sons, Inc.

ISBN: 9781843393238. Published by IWA Publishing, London, UK.

Birks, R., Colbourne, J., Hills, S., and Hobson, R. (2004). Microbiological water quality in a large building water recycling facility. *Water Science and Technology*, **50**(2), 165–172.

Biswas, A.K., Khoshoo, T., and Khosla, A. (1990). *Environmental Modelling for Developing Countries*. London: Tycooly Publishing.

Bocas.org. (2009, September 2). *Bocas del Toro*. Retrieved September 2, 2009, from http://www.bocas.org/maps.php.

Brown, D.E., Fo, M., and Pelletier, M.R. (Eds). (2000). *Sustainable Architecture White Pages*. New York: Earth Pledge Foundation.

Bruijn, M.D., van Halsema, I., and van den Hombergh, H. (1997). *Gender and Land Use: Diversity in Environmental Practices*. Amsterdam: Thela Publishers.

Camberlin, P. and Wairoto, J. (1997). Intraseasonal wind anomalies related to wet and dry spells during the "Long" and "Short" rainy seasons in Kenya. *Theoretical and Applied Climatology*, **58**(1–2).

Cardoze, N.M. and Tempone, C. (1985). *Guide for Research Workers of Panama*. Panama: Pan American Institute of Geography and History.

Centre for Science and Environment. (2005a). *Components of a Rainwater Harvesting System*. Retrieved December 6, 2005, from http://www.rainwaterharvesting.org/Urban/Components.htm.

Centre for Science and Environment. (2005b). *Legislation on Rainwater Harvesting*. Retrieved December 6, 2005, from http://www.rainwaterharvesting.org/Urban/Legislation.htm.

CIA. (2009). *Central Intelligence Agency*. Retrieved March 25, 2006, from The World Factbook: https://www.cia.gov/library/publications/the-world-factbook/geos/ke.html.

Clivus. (2002). *Composter Models*. Retrieved 2005, from http://www.clivusne.com/Vsm/templates/temp.asp?articleid=92andzoneid=35.

De, S. and Ariyabandu, R. (2003, April). *Very Low-Cost Domestic Roofwater Harvesting in the Humid Tropics: Its Role in Water Policy*. Retrieved February 2006, from http://www.eng.warwick.ac.uk/DTU/rwh/dfiddocs/r4.pdf.

Deago, E. (2005). *Environmental Impact Study Category I. Treatment System and Final Disposal of Wastewater for Smithsonian Tropical Research Institute on Isla Colón.*

Dixon, A., Butler, D., and Fewkes, A. (1999). Water saving potential of domestic water reuse systems using greywater and rainwater combination. *Water Science and Technology*, **39**(5), 25–32.

DTU. (2003). *Development Technology Unit, University of Warwick School of Engineering*. Retrieved November 11, 2005, from Roofwater Harvesting for Poorer Households in the Tropics: http://www.eng.warwick.ac.uk/DTU/rwh/dfid.html.

DTU. (2005b). *Development Technology Unit, University of Warwick School of Engineering*. Retrieved September 2009, from Filters and Separators: http://www2.warwick.ac.uk/fac/sci/eng/research/dtu/rwh/technology1/technology_5/.

DTU. (2005c). *Development Technology Unit, University of Warwick School of Engineering*. Retrieved September 26, 2009, from Gutters: http://www2.warwick.ac.uk/fac/sci/eng/research/dtu/rwh/technology1/technology_3/.

DTU. (2005e). *Development Technology Unit, University of Warwick School of Engineering*. Retrieved December 26, 2005, from Sizing a DRWH System: http://www.eng.warwick.ac.uk/DTU/rwh/sizing.html.

DTU. (2005f). *Development Technology Unit, University of Warwick School of Engineering*. Retrieved December 26, 2005, from Styles of Roofwater Harvesting: http://www.eng.warwick.ac.uk/DTU/rwh/styles.html.

EAWAG. (2005). *Solar Water Disinfection-How does it work?* Retrieved February 24, 2006, from The Swiss Federal Institute for Environmental Science and Technology: http://www.sodis.ch/Text2002/T-Howdoesitwork.htm.

European Commission. (2004). *Drinking Water Directive*. Retrieved from Synthesis Report: http://ec.europa.eu/environment/water/water-drink/index_en.html.

Falkenmark, M. (2007). Shift in thinking to address the 21st century hunger gap: Moving focus from blue to green water management. *Water Resources Management*, **21**, 3–18.

Government of Canada. (2008, March). *Water and Climate Change, and the Energy-Water Nexus*. Retrieved September 2009, from The Government of Canada: Policy Research Initiative. The Science-Policy Interface: http://www.policyresearch.gc.ca/doclib/SR_ENG_Science-Nexus_200803_e.pdf.

Gupta, A. (1998). *Ecology and Development in the Third World*. New York: Routledge.

Guzha, E. and Musara, C. (2003, April). *Assessment of Community Knowledge, Attitudes, Practice, Behaviour and Acceptance of Ecological Sanitation in Peri-Urban Areas of Harare*. Retrieved 2005, from Second International Symposium on Ecological Sanitation: http://www2.gtz.de/ecosan/download/ecosan-Symposium-Luebeck-session-b.pdf.

Hach. (2003). Hach Booklet, Hach User's Manual for Hach DR/890 Colorimeter.

Hach. (2005). *The Hach Company*. Retrieved December 1, 2005, from DR/890 Colorimeter Procedures Manual: http://www.hach.com/fmmimghach?/CODE%3A4847022_7ED9083%7C1.

Handia, L. (2005). Comparative study of rainwater quality in urban Zambia. *Aqua*, 55–64.

Hannan, C. and Andersson, I. (2002). Retrieved August 24, 2005, from Gender Perspectives on Water Supply and Sanitation: Towards a Sustainable Livelihoods and Ecosystem-Based Approach to Sanitation: www.undp.org/water/docs/sum_aag_pre.doc.

Hoekstra, A. and Chapagain, A. (2007). Water footprints of nations: Water use by people as a function of their consumption pattern. *Water Resources Management*, **21**, 35–48.

Holden, R. (2002). *The Use of Dry Sanitation in an Urban Environment*. Mvula Trust Case Study Series Number 7.

Hoyle, B. (2008, April). *The energy-water nexus: deja-vu all over again?* Retrieved September 2009, from Nature Reports Climate Change: http://www.nature.com/climate/2008/0804/full/climate.2008.22.html.

Hunnan, C. and Andersson, I. (2002). *Gender Perspectives on Ecological Sanitation*. Retrieved August 2005, from http://www2.gtz.de/ecosan/download/Hannan-gender-perspectives.pdf.

IDE. (2004). *International Development Enterprises*. Retrieved November 11, 2005, from Fostering economic independence for fellow farmers, Case study: Rajesh Patil: http://www.ide-india.org/ide/cs1-rajesh.shtml.

IEA and SID. (2001). *Kenya at the Crossroads: Scenarios for Our Future*. Nairobi: Institute of Economic Affairs (IEA) and Society for International Development (SID).

IFRC. (2006). *Kenya: Drought as of 4 Jan 2006.* Retrieved April 8, 2006, from ReliefWeb: http://www.reliefweb.int/rw/RWB.NSF/db900SID/AHAA-6KQRGJ?OpenDocumentandrc=1andemid=ACOS-635NZE.

Kennedy, J. (Ed.). (2004). *Building without Borders: Sustainable Construction for the Global Village.* Gabriola Island, Canada: New Society Publishers.

Kenya Weather. (n.d.). Retrieved August 24, 2005, from http://www.kenya-travel-safari.com/weather.html.

Khor, M. and Lin, L.L. (2001a). *Good Practices and Innovative Experiences in the South: Economic, Environmental and Sustainable Livelihood Initiatives* (Vol. 1). New York: Zed Books.

Khor, M. and Lin, L.L. (2001b). *Good Practices and Innovative Experiences in the South: Social Policies, Indigenous Knowledge and Appropriate Technology* (Vol. 2). New York: Zed Books.

Kirby, A. (2002, April). *Water Key to Ending Africa's Poverty.* Retrieved March 26, 2006, from BBC: http://news.bbc.co.uk/1/hi/sci/tech/1920789.stm.

Laikipia Wildlife Forum. (2009). Retrieved August 24, 2005, from http://www.laikipia.org.

LaMotte. (2005). *ColiQuant EZ.* Retrieved 2009, from http://www.uwex.edu/ces/csreesvolmon/EColi/LamotteEZgelInfo.pdf.

LaMotte. (2009, September). *Bacteria Studies.* Retrieved September 2009, from http://www.lamotte.com/pages/edu/3-0034.html.

LaMotte. (2009, September). *pH (Individual Test Kits).* Retrieved from http://www.lamotte.com/pages/wawa/ind-kits/ph-kits.html.

Lozán, J.L., Grassl, H., Hupfer, P., Menzel, L., and Schönwiese, C.-D. (2004). The water problem of our earth: from climate and the water cycle to the human right for water. In *Global Change: Enough Water for All?* Hamburg: PIK Publications.

Macomber, P. (2001). *Guidelines on Rainwater Catchment Systems for Hawaii.* Retrieved 2006, from College of Tropical Agriculture and Human Resources, University of Hawaii at Manoa: http://www.ctahr.hawaii.edu/oc/freepubs/pdf/RM-12.pdf.

Mallinckrodt Baker. (2006). *Product Catalog.* Retrieved September 2009, from BAKER-pHIX pH Indicator Sticks: http://www.mallbaker.com/americas/catalog/prodpage.asp?brd=bandsku=4390andmarketname=andmarkethead=andsearchda-andmarkethead=andsearchdata=4390andp%20um=0andmaxitem=1.

Martinson, D.B. and Thomas, T. (2003, August). *Improving Water Quality by Design.* Retrieved January 13, 2006, from UK Department for International Development: http://www.eng.warwick.ac.uk/DTU/pubs/outside/ircsa11wqbydesign.pdf.

Mbaru, J. (2003). *Transforming Africa: New Pathways to Development.* Nairobi: East African Educational Publishers.

Mead, S.P. (2001, April). *Green Building: Current Status and Implications for Construction Education.* Retrieved September 24, 2005, from ASC Proceedings of the 37th Annual Conference: http://asceditor.unl.edu/archives/2001/mead01.htm.

Mendler, S. and Odell, W. (2000). *The HOK Guidebook to Sustainable Design.* New York: John Wiley and Sons, Inc.

Micron. (2005). *Ross Controls.* Retrieved December 19, 2005, from http://www.rosscontrols.com/micron.htm.

Millea, N. (2002). *Big and Green.* New York: Princeton Architectural Press.

Ministry of Commerce and Industry of Panama. (1999, December). *Technical Rule DGNT-COPANIT 23-395-99*. Retrieved February 12, 2006, from http://www.ersp.gob.pa/agua/Anexos/395%20Gaceta%20Oficial.pdf.
Ministry of Commerce and Industry of Panama. (2000). *DGNTI-COPANIT 47-2000*. Retrieved 2005, from http://www.cnpml.org.pa/cnpml/leyes_normas/copanit47_2000_lodos.pdf.
Ministry of Water and Irrigation Laikipia District. (2005). *Physical/Chemical and Microbiological Examination of Recreational Waters*. Ministry of Water and Irrigation Laikipia District, Kenya.
Morgan, P. (2002). *Ecosanres*. Retrieved August 2005, from An Ecological Approach to Sanitation in Africa: A Compilation of Experiences: http://www.ecosanres.org/PM_Report.htm.
Mpala Wildlife Foundation. (2003). *Environment and Geography*. Retrieved October 2005, from Mpala Wildlife Foundation: http://www.mpala.org/researchctr/environment/index.html.
Mpala Wildlife Foundation. (2005). *Area Maps*. Retrieved October 20, 2005, from Mpala Wildlife Foundation: http://www.mpala.org/researchctr/environment/map.html.
Mudavadi, M. (2002). *Rethinking Development in Africa in the 21st Century: The Case of Kenya*. Nairobi: East African Educational Publishers.
Mwagore, D. (2002). *Land Use in Kenya: the Case for a National Land Use Policy*. Retrieved December 2, 2005, from Kenya Land Alliance: http://www.oxfam.org.uk/what_we_do/issues/livelihoods/landrights/downloads/klasmall.pdf.
Nagel, S.S. (2002). *Environmental Policy and Developing Nations*. Jefferson, North Carolina: McFarland and Co., Inc.
North Dakota State University. (1992). *Activated Carbon Filtration*. Retrieved December 19, 2005, from http://www.ext.nodak.edu/extpubs/h2oqual/watsys/ae1029w.htm.
Odhiambo, C., Petters, J., and Wreford, K. (2005). *Water Supply and Consumption at the Mpala Conservancy*. The Mpala Research Centre.
Onyenechere, E. (2004). Water Supply Measures Used by the Rural People of Ebonyi and Enugu, Nigeria. *Journal of Water Supply Research and Technology*, **53**, 425–431.
Pacey, A. and Cullis, A. (1986). *Rainwater harvesting: The Collection of Rainfall and Run-off in Rural Areas*. London: Intermediate Technology Publications.
Panikkar, A.A. (2003). *Issues in Blackwater Reuse: An Example of a Treatment System*. The University of Western Sydney, Australia: School of Engineering and Industrial Design.
Pascal, F. (2000). *Initial Site Assessment Report. Memo from Smithsonian.*
Petters, J. (2006, January 24). (M. Waite, Interviewer).
Rajan, M.G. (1997). *Global Environmental Politics: India and the North-South Politics of Global Environmental Issues*. New Delhi: Oxford University Press.
Rathgeber, E. (2005). *Women, Men and Water-Resource Management in Africa*. Retrieved February 12, 2006, from International Development Research Centre (IDRC): http://www.idrc.ca/en/ev-31108-201-1-DO_TOPIC.html.
ReliefWeb. (2003a). *Kenya: Cumulative Rainfall Oct-Dec 2003, Short Rains Compared to Normal*. Retrieved April 7, 2006, from http://www.reliefweb.int/rw/rwb.nsf/db900SID/SKAR-64GE3Q?OpenDocumentandemid=ACOS-635NZEandrc=1.
ReliefWeb. (2003b). *Kenya: Rainfall Estimates (20 Mar–10 Jun 2003) Vs. Average*. Retrieved April 7, 2006, from http://www.reliefweb.int/rw/rwb.nsf/db900SID/SKAR-64GDXP?OpenDocumentandemid=ACOS-635NZEandrc=1.

ReliefWeb. (2004). *Kenya Drought-Situation Map.* Retrieved April 5, 2006, from http://www.reliefweb.int/rw/RWB.NSF/db900SID/SKAR-64GE5P?OpenDocument andrc=1andemid=ACOS-635NZE.

ReliefWeb. (2005, July). *Kenya: Thousands Displaced After Attack, Toll Reaches 76.* Retrieved April 5, 2006, from http://www.reliefweb.int/rw/fullMaps_Af.nsf/luFullMap/E2B590A85198074985257O3E00798AC9/$File/rw_ce_ken140705.pdf?OpenElement.

Rooney, R. (2005, December 2). Environmental Health Advisor for WHO. (M. Waite, Interviewer).

Saavedra, C. (2005, October 31). (M. Waite, Interviewer) personal interview on date stated.

SACG. (2005, May). *Housing.* Retrieved October 15, 2005, from South African Consulate General: http://www.southafrica-newyork.net/consulate/housing.htm.

Safe Water Systems. (2005). *What is Solar Water Pasteurization?* Retrieved January 2, 2006, from http://o.b5z.net/i/u/10024191/f/Literature/What%20is%20Solar%20Water%20Pasteurization.pdf.

Sanitron. (2005). *Sanitron Ultraviolet Water Purifiers.* Retrieved December 18, 2005, from Atlantic Ultraviolet Corporation: http://www.ultraviolet.com/pdflib/981169.pdf

Schnoor, J. (1997). *Phytoremediation: Groundwater Remediation Technologies Analysis.* Center Technology Evaluation Report, TE-98-01, 37.

Sedelmeier, H. (2005, October 31). (M. Waite, Interviewer) personal interview on date stated.

Shuttleworth, W.J. (2007). Stakeholder-driven, enquiry-driven, or stakeholder-relevant, enquiry-driven science? *Water Resources Management*, **21**, 63–77.

Smith, F. (1997). *Environmental Sustainability: Practical Global Implications.* Boca Raton, Florida: St. Lucie Press.

Stang, A. and Hawthorne, C. (2005). *The Green House: New Directions in Sustainable Architecture.* New York: Princeton Architectural Press.

STRI. (2000). *Characterization and Environmental Plan Marine Research Facilities Bocas del Toro, Panama.* Bocas del Toro: Smithsonian Tropical Research Institute.

STRI. (2005). *U.S. Department of Energy High Performance Federal Buildings.* Retrieved from Smithsonian Tropical Research Institute: http://femp.buildinggreen.com/overview.cfm?ProjectID=263.

STRI. (2006). *Mpala RC Downloads.* Retrieved April 4, 2006, from Smithsonian Tropical Research Institute: http://striweb.si.edu/esp/physical_monitoring/download_mpala.htm.

Thomas, P. (1994). Macrophyte Trenches for Septic Tank Effluent. *Affordable Water Supply and Sanitation.* 20th Water Engineering Development Centre (WEDC) Conference.

Thomas, P. and Greene, G. (1993). Rainwater Quality from Different Roof Catchments. *Water, Science and Technology*, **28**(3–5), 291–299.

Thomas, T. and Martinson, D. (2007). *Roofwater Harvesting: A Handbook for Practitioners.* Retrieved September 26, 2009, from http://www.irc.nl/page/37471.

Turner, S. (2000, April). *Design of Rainwater Storage Tanks for use in Developing Countries.* Retrieved December 2005, from University of Warwick School of Engineering: http://www.cd3wd.com/cd3wd_40/UWWKDTU/ugp001_turner/PDF/ugp001_turner.pdf.

UNEP. (1999). *Sourcebook of Alternative Technologies for Freshwater Augmentation in Small Island Developing States.* Retrieved December 9, 2005, from United Nations Environment Programme: http://www.unep.or.jp/ietc/Publications/TechPublications/TechPub-8d/harvesting.asp.

UNEP. (2002a, March). *International Source Book on Environmentally Sound Technologies for Wastewater and Stormwater Management.* Retrieved November 27, 2005, from United Nations Environment Programme: http://www.unep.or.jp/ietc/Publications/TechPublications/TechPub-15/main_index.asp.

UNEP. (2002b). *Rainwater Harvesting and Utilisation.* Retrieved December 10, 2005, from United Nations Environment Programme: http://www.unep.or.jp/Ietc/Publications/Urban/UrbanEnv-2/index.asp.

UNEP. (2003). *GEO Year Book 2003.* Retrieved 2005, from Land use: a critical factor: http://www.unep.org/geo/yearbook/yb2003/fig40.htm.

UNEP. (2003, March). *Phytotechnologies: A Technical Approach in Environmental Management.* Retrieved November 24, 2005, from United Nations Environment Programme: http://www.unep.or.jp/ietc/Publications/Freshwater/FMS7/index.asp.

UNDP. (2005). *Ecological Sanitation.* Retrieved November 11, 2005, from United Nations Development Programme: http://www.undp.org/water/ecol.html.

UNESCO. (2003, March). *Water for People Water for Life.* Retrieved September 2009, from The United Nations World Water Development Report: http://unesdoc.unesco.org/images/0012/001295/129556e.pdf.

UNESCO-WWAP. (2009). *Climate Change and Water–An Overview from the World Water Development Report 3: Water in a Changing World.* Retrieved December 2009, from A United Nations World Water Assessment Programme Special Report: http://www.unesco.org/water/wwap/wwdr/wwdr3/pdf/WWDR3_Water_in_a_Changing_World.pdf.

UN-Habitat. (2001, June). *Urbanization Facts and Figures.* Retrieved from http://www.unhabitat.org/istanbul+5/bg10.htm.

UNICEF. (2002a). *At a glance: Kenya.* Retrieved January 13, 2006, from United Nations Children Fund: http://www.unicef.org/infobycountry/kenya_statistics.html.

UNICEF. (2002b). *At a glance: Panama.* Retrieved January 13, 2006, from United Nations Children Fund: http://www.unicef.org/infobycountry/panama_statistics.html.

University of Hong Kong. (2005). *Sustainable Architecture.* Retrieved October 23, 2005, from The University of Hong Kong Building Energy Efficiency Research: http://www.arch.hku.hk/research/BEER/sustain.htm#3.3.

University of Warwick. (2007, December 26). *Introduction to domestic roofwater harvesting.* Retrieved February 2006, from Development Technology Unit: http://www2.warwick.ac.uk/fac/sci/eng/research/dtu/rwh/.

USDOE. (2004, July 30). *Smithsonian Tropical Research Institute.* Retrieved 2005, from US Department of Energy: http://femp.buildinggreen.com/overview.cfm?ProjectID=263.

USEPA. (2005, November). (M. Waite, Interviewer) phone interview USEPA hotline (names are not allowed to be given).

USEPA. (2005a). *United States Environmental Protection Agency.* Retrieved December 19, 2005, from Drinking Water Contaminants: http://www.epa.gov/safewater/hfacts.html#Synthetic.

USEPA. (2005b). *United States Environmental Protection agency*. Retrieved November 28, 2005, from Municipal Technologies: http://www.epa.gov/owm/mtb/mtbfact.htm.

USEPA. (2009, August). *Drinking Water Contaminants*. Retrieved November 23, 2005, from http://www.epa.gov/safewater/mcl.html#4.

USGBC. (2005, October). *United States Green Building Council*. Retrieved from Why Build Green? : http://www.usgbc.org/DisplayPage.aspx?CMSPageID=38&.

Villagra, J. (2005, November 1). (M. Waite, Interviewer) personal interview on date stated.

Vries, F.W. (2007). Learning Alliances for the broad implementation of an integrated approach to multiple sources, multiple uses and multiple users of water. *Water Resources Management*, **21**, 79–95.

Waite, M. (2005). Wastewater recycling and ecological sanitation: Technologies, methods, and approaches. *Unpublished Manuscript*. Water Supply and Sanitation Collaborative Council (WSSCC).

Water Systems Council. (2004, July). *pH in Drinking Water*. Retrieved December 15, 2005, from www.watersystemscouncil.org/VAiWebDocs/WSCDocs/3531585PH.PDF.

WHO. (1997). *Guidelines for drinking water quality*. Retrieved 2005, from Volume 3, Surveillance and control of community supplies: http://www.who.int/water_sanitation_health/dwq/gdwqvol32ed.pdf.

WHO. (2005). *Annex 4 Chemical summary tables*. Retrieved 2005, from http://www.who.int/water_sanitation_health/dwq/en/gdwq3_ann4tab.pdf.

WHO. (2006). *Guidelines for drinking-water quality, third edition, incorporating first and second addenda*. Retrieved 2006, from http://www.who.int/water_sanitation_health/dwq/gdwq3rev/en/.

WHO and UNICEF JMP. (2005). *Guide for Drinking Water Supply, Sanitation and Hygiene Related Household Survey Questions (DRAFT)*. Retrieved from http://whqlibdoc.who.int/publications/2006/9789241563260_eng.pdf.

Winbald, U. (2004). *The Next Generation Toilet: A Global Perspective*. Stockholm: Stockholm Environment Institute.

Yeang, K. (1995). *Designing with Nature: The Ecological Basis for Architectural Design*. New York: McGraw-Hill, Inc.

Zeiher, L.C. (1996). *The Ecology of Architecture: A Complete Guide to Creating the Environmentally Conscious Building*. New York: Watson-Guptill Publications.

Zhu, Q. and Li, Y. (2004). Rainwater Harvesting: An Alternative for Securing Food Production under Climate Variability. *Water Science and Technology*, **49**(7), 157–163.

11
Appendix

©2010 IWA Publishing. *Sustainable Water Resources in the Built Environment*. By Marilyn Waite.
ISBN: 9781843393238. Published by IWA Publishing, London, UK.

Materials for UV Water Disinfection

Quantity	Description
1	Ultraviolet water purifier 40GPM, 120V/60Hz 316SST, Sanitron, model S2400C (replaces model #S2400B)
1	Monitor: 118V/60Hz for S2400 up, S2400/A2400 complete with probe
1	Audio alarm: 118V/60Hz complete here
1	2" NPT, burkert solenoid valve: 120V/60Hz, 5281SB-A50
1	40 GPM, 2" NPT – PVC (black)
1	Time delay mechanism
1	Elapsed time ind: remote 118V/60Hz complete assy
1	Filter chamber,model D 2412 up to 2400 GPH, filter corrosion resistant filter chamber, 2" inlet/outlet w/12 to 20 filter tube chamber
1	Pressure tank 85 gls
1	Pressure switch 30/50 psi
1	Control level
2	Check valve, 2"piston
1	Manometer c/glycerin
3	Water reserve tanks, 1,000 gls black
5	Codos de cobre 2x90°
1	Copper Tee 2"
3	Welding valves 2 Nibco SI8I N00140D

Figures A.1 UV Water disinfection system details for STRI building in Panama

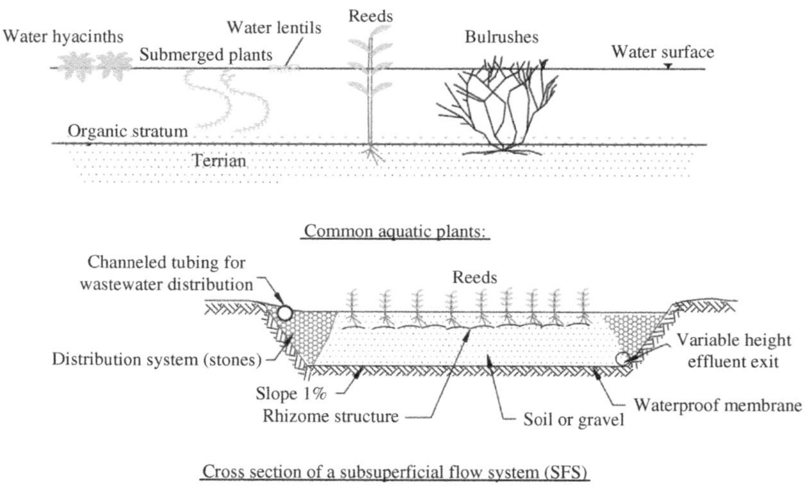

Figure A.2 Native plants for constructed wetland at STRI building in Panama

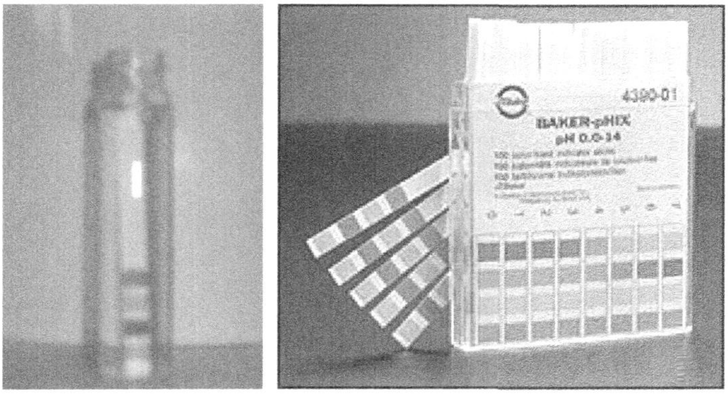

Figure A.3 pH Experiment (Baker-pHix pH Indicator Stick Kit)

Figure A.4 Balance

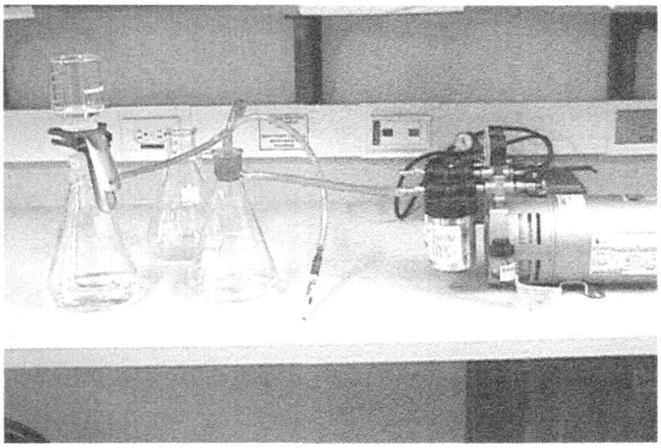

Figure A.5 Filtration with suction pump

Figure A.6 Desiccator

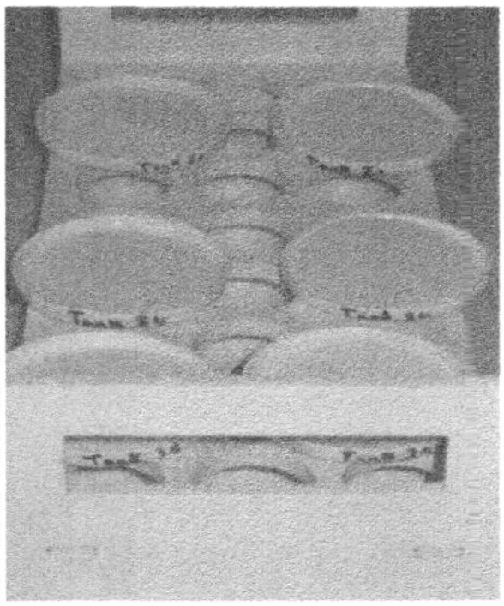

Filter paper for various water samples

Figures A.4–A.6 Total suspended solids experiment completed in Panama

Figure A.7 Experiments using Hach DR/890 Colorimeter

Figure A.8 Coliform experiment completed in Panama

Table A.1 TSS procedures

Step Number	Procedure Description
1	A dried filter paper was weighed on a balance to the nearest 0.001 g.
2	This weight was recorded as 'B' for later calculations.
3	100 mL of sample was measured.
4	The pre-weighed filter paper was placed on the filtration apparatus.
5	The sample was poured into the filtration apparatus and a vacuum was applied.
6	Suction was continued in order to remove all traces of water, and then the vacuum was turned off.
7	The filter paper was removed and placed onto oven-safe containers.
8	The filter paper containers were placed in the oven at 103–105°C for 1 hour.
9	The containers were removed from the oven and allowed to cool in a desiccator.
10	After the filter paper cooled, it was weighed and recorded as 'A'.

The following calculation was performed to determine TSS:

$$\frac{mg\ total\ suspended\ solids}{L} = \frac{(A - B) \times 1000}{Sample\ volume\ (mL)}$$

A = weight of filter + dried residue, mg
B = weight of filter, mg

Table A.2 Coliform procedures (LaMotte, 2005)

Step Number	Procedure Description
1	A permanent marker was used to label a petri dish and a bottle of Coliscan Easygel with the sample information. The dish was taped shut with two pieces of clear tape. One piece of tape was used as the hinge and the other as a latch.
2	The cap was removed from one bottle of Coliscan Easygel, and a 1mL pipet was used to pipet 5 mL of the sample into the bottle.
3	The bottle was capped and swirled to distribute the contents.
4	The tape was removed from one side of the petri dish and the lid was lifted. The mixture was poured into the petri dish bottom.
5	The lid on the petri dish was then replaced. The dish was carefully swirled until the bottom of the dish was completely covered with liquid.
6	The dish was taped shut.
7	While the contents of the dish were still liquid, the dish was placed with the lid side up in an incubator. The liquid solidified in about 30–45 minutes.
8	When the medium had solidified, the dish was turned upside down (with the lid side down).
9	The Petri dish was incubated for 48–72 hours in a preheated incubator at 95°F (35–37°C).
10	Colonies began to appear after 24 hours.
11	The colonies were counted with the dish upside down. Colonies are were reported as Colony Forming Units (CFU) per 100 milliliter of water.
12	To determine the result as E. Coli or Fecal Coliform, the dark blue to purple colonies were counted. Light-blue, blue-green or white colonies were disregarded.

- To report the result in terms of E. coli or Fecal Coliform per mL of water, the number of colonies were divided by the sample size:

$$E.\ Coli\ or\ Fecal\ Coliform\ Colonies\ per\ mL\left(\frac{CFU}{mL}\right)=\frac{\#\ Colonies}{Amount\ of\ sample\ used}$$

- To report the results in terms of E. Coli or Fecal Coliform per 100 mL of water, the number of colonies were multiplied by 100 and then divided by the sample size:

$$E.\ Coli\ or\ Fecal\ Coliform\ Colonies\ per\ 100\ mL\left(\frac{CFU}{100\ mL}\right)=\frac{\#\ Colonies\times 100}{Amount\ of\ sample\ used}$$

To determine the result as Total Coliforms (E. Coli + General coliforms), the number of pink and purple colonies were counted.

- To report the result in terms of Total Coliforms per mL of water: the number of colonies were divided by the sample size

$$Total\ Coliform\ Colonies\ per\ mL\left(\frac{CFU}{mL}\right)=\frac{\#\ Colonies}{Amount\ of\ sample\ used}$$

- To report the results in terms of Total Coliforms per 100 mL of water, the numbers of colonies were multiplied by 100, and then divide by the sample size.

$$Total\ Coliform\ Colonies\ per\ 100\ mL\left(\frac{CFU}{100\ mL}\right)=\frac{\#\ Colonies\times 100}{Amount\ of\ sample\ used}$$

Table A.3 Method for true and apparent color tests

Step Number	Procedure Description
1	The filtering apparatus was assembled (membrane filter, filter holder, filter flask, and aspirator). Note: To test for apparent color filtering was not performed.
2	The filter was rinsed by pouring about 50 mL of deionized water through the filter and discarding the rinse water.
3	Another 50 mL of deionized water was poured through the filter.
4	A sample cell (the blank) was filled with 25 mL of filtered deionized water (from Step 3). Note: For apparent color, unfiltered deionized water was used.
5	The PRGM 19 was entered. The display will showed Pt-Co and the ZERO icon.
6	About 50 mL of the sample was poured through the filter.
7	A second sample cell (the prepared sample) was filled with 25 mL of the filtered sample.
8	The blank was placed into the cell holder and covered with the instrument cap.
9	ZERO was pressed, the cursor moved to the right, then the display showed 0 mg/L Pt-Co.
10	The prepared sample was placed into the cell holder and covered with the instrument cap.
11	READ was pressed. The cursor moved to the right, and then the result in Platinum-Cobalt color units (Pt-Co) was displayed.

Table A.4 Method for nitrate tests

Step Number	Procedure Description
1	The PRGM 51 ENTER was pressed on the colorimeter.
2	A sample cell was filled with 10 mL of sample.
3	The contents of one NitraVer 5 Nitrate Reagent Powder Pillow were added to the sample cell (the prepared sample), and the sample cell was capped.
4	TIMER ENTER was pressed to initiate a one-minute reaction period. During this period, the sample cell was shook vigorously until the timer beeped.
5	After the timer beeped, the display showed 5:00 TIMER.
6	ENTER was pressed to allow a five-minute reaction period to begin. Note: an amber color would develop if nitrate nitrogen was present.
7	Another cell was filled with 10 mL of sample (the blank).
8	The blank was placed into the cell holder and covered with the instrument cap.
9	When the timer beeped, ZERO was pressed. The cursor moved to the right, and then the display showed 0.0 mg/L NO3-N.
10	The prepared sample was placed into the cell holder and covered with the instrument cap.
11	READ was pressed, the cursor moved to the right, and then the result in mg/L NO3-N was displayed.

Table A.5 Method for nitrite tests

Step Number	Procedure Description
1	The PRGM 59 was entered on the colorimeter.
2	A sample cell was filled with 10 mL of sample.
3	The contents of one NitriVer 2 Nitrite Reagent Powder Pillow were added to the cell and inverted 5–7 times to mix (the prepared sample). Note: A greenish-brown color would develop if nitrite was present.
4	TIMER ENTER was pressed to allow a ten-minute reaction period to begin.
5	Another sample cell was filled with 10 mL of sample (the blank).
6	The blank was placed into the cell holder and covered with the instrument cap.
7	ZERO was pressed. The cursor moved to the right, then the display showed: 0 mg/L NO2
8	After the timer beeped, the sample was gently inverted twice. It was then placed into the cell holder and covered with the instrument cap.
9	READ was pressed. The cursor moved to the right, and then the result in mg/L nitrite was displayed.

Table A.6 Method for sulfate tests

Step Number	Procedure Description
1	The PRGM 91 was entered on the colorimeter.
2	A sample cell was filled with 10 mL of sample.
3	The contents of one SulfaVer 4 Sulfate Reagent Powder Pillow were added to the sample cell (the prepared sample). The cell was capped and inverted several times to mix. Note: A white turbidity would develop if sulfate was present in the sample. Accuracy was not affected by any undissolved powder.
4	TIMER ENTER was pressed so that a 5-minute reaction period could begin. The cell was allowed to stand undisturbed.
5	After the timer beeped, a second sample cell was filled with 10 mL of sample (the blank).
6	The blank was placed into the cell holder.
7	ZERO was pressed. The cursor moved to the right, and then the display showed 0 mg/L SO4.
8	Within five minutes after the timer beeped, the prepared sample was placed into the cell holder.
9	READ was pressed, and the cursor moved to the right to display the result in mg/L SO4.

Table A.7 Method for turbidity tests

Step Number	Procedure Description
1	The PRGM 95 was entered on the colorimeter (1 FAU = 1 NTU = 1 FTU when measuring formazin).
2	A sample cell was filled with 10 mL of deionized water (the blank).
3	The blank was placed into the cell holder and covered with the colorimeter cap.
4	ZERO was pressed on the colorimeter. The cursor moved to the right, and then the display showed: 0 FAU.
5	Another sample cell was filled with 10 mL of sample.
6	The sample cell was placed into the cell holder and tightly covered with the instrument cap.
7	READ was pressed on the colorimeter. The cursor moved to the right, and then the result in Formazin Attenuation Units (FAU) was displayed.

ColiQuantz EZ Colony Color Guide

The left half of the photo (1–4) consists of colonies of *E.coli* (1A, 1B) and *Enterobacter aerogenes* (2, 3, 4) growing in/on Coliscan® Easygel® medium. The right half of the photo (5, 6, 7) represents the appearance of organisms other than *E. coli* or coliforms.

(1A)
Two *E. coli* CFUs showing purple color with obvious pink diffused halos. Fecal coliform (Glucuronidase+, Galactosidase+)

(1B)
E. coli CFUs blue/purple color with minimal pink halos. Fecal coliform (Glucuronidase+, Galactosidase+)

(7)
Colorless CFU (indicates no Glucuronidase or Galactosidase activity). This colony type should not be counted as *E.coli* or coliform.*

(5)
Teal green CFU growing on surface of medium. (Glucuronidase+) This colony type should not be counted as *E.coli* or coliform.*

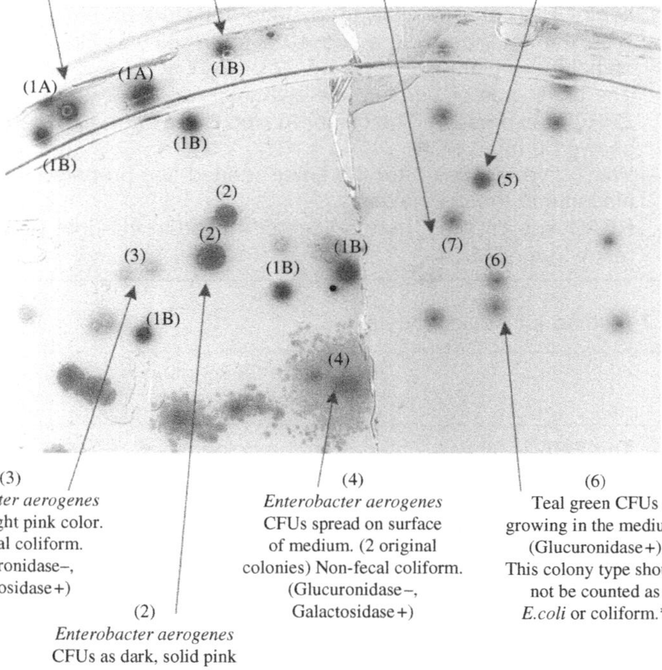

(3)
Enterobacter aerogenes CFUs as light pink color. Non-fecal coliform. (Glucuronidase–, Galactosidase+)

(2)
Enterobacter aerogenes CFUs as dark, solid pink color. Non-fecal coliform. (Contain no blue/purple) (Glucuronidase–, Galactosidase+)

(4)
Enterobacter aerogenes CFUs spread on surface of medium. (2 original colonies) Non-fecal coliform. (Glucuronidase–, Galactosidase+)

(6)
Teal green CFUs growing in the medium. (Glucuronidase+) This colony type should not be counted as *E.coli* or coliform.*

*These teal or colorless types of colonies may be significant other types of bacteria (such as Salmonella spp. or Shigella spp.) or even rarely found atypical E. coli or coliforms, but should never be counted as *E.coli* or coliform without further biochemical testing
Photo and information for Color Guide supplied by Micrology Laboratories, LLC.

LaMatte Company • Chestertown • MD • 21620 • 800-344-3100 • www.lamotte.com

Figure A.9 Colony color guide for total and fecal coliform experiments completed in Kenya and Panama

Table A.8 Calculations for coliform tests completed in Panama

Source	Trial #/ Storage Tank #	# of Fecal Colonies (dark blue to purple)	# of Non-Fecal Colonies (pink)	Fecal Coliform (CFU/100mL)	Total Coliform (CFU/100mL)
Municipal Water	**Trial 1**				
(from Sink at STRI lab)	Petri Dish 1	0	0	**0**	**0**
	Petri Dish 2	0	0	**0**	**0**
Rain from Roof	**Trial 1**				
	Petri Dish 1	0	7	**0**	**140**
	Petri Dish 2	0	12	**0**	**240**
	Trial 2				
	Petri Dish 1	33	10	**660**	**860**
	Petri Dish 2	19	0	**380**	**380**
Rain Stored in Rainwater	**Storage Tank 1**				
Storage Tanks	Petri Dish 1	1	21	**20**	**440**
	Petri Dish 2	6	16	**120**	**440**
	Storage Tank 2				
	Petri Dish 1	13	10	**260**	**460**
	Petri Dish 2	4	21	**80**	**500**
	Storage Tank 3				
	Petri Dish 1	14	2	**280**	**320**
	Petri Dish 2	4	10	**80**	**280**
Rain from Overflow Pipe	**Trial 1**				
	Petri Dish 1	22	10	**440**	**640**
	Petri Dish 2	0	48	**0**	**960**

Table A.9 Statistical analysis for TSS and coliform tests in Panama

TSS for Rain Stored in Rainwater Storage Tanks

Mean	**17.78**
Standard Error	5.21
Median	20
Mode	20
Standard Deviation	**15.63**
Sample Variance	244.44
Kurtosis	1.29
Skewness	0.97
Range	50
Minimum	**0**
Maximum	**50**
Sum	160
Count	9

TSS for Rainwater from Roof

Mean	**35**
Standard Error	4.28
Median	35
Mode	40
Standard Deviation	**10.49**
Sample Variance	110
Kurtosis	−0.25
Skewness	3.747E-17
Range	30
Minimum	**20**
Maximum	**50**
Sum	210
Count	6

Fecal Coliform for Rain Stored in Rainwater Storage Tanks

Mean	**140**
Standard Error	43.20
Median	100
Mode	80
Standard Deviation	**105.83**
Sample Variance	11200
Kurtosis	−1.65
Skewness	0.58
Range	260
Minimum	**20**
Maximum	**280**
Sum	840
Count	6

Table A.9 *Continued*

Total Coliform for Rain Stored in Rainwater Storage Tanks

Mean	**406.67**
Standard Error	35.28
Median	440
Mode	440
Standard Deviation	**86.41**
Sample Variance	7466.67
Kurtosis	−1.22
Skewness	−0.76
Range	220
Minimum	**280**
Maximum	**500**
Sum	2440
Count	6

Fecal Coliform for Rainwater from Roof

Mean	**260**
Standard Error	160.62
Median	190
Mode	0
Standard Deviation	**321.25**
Sample Variance	103200
Kurtosis	−2.56
Skewness	0.61
Range	660
Minimum	**0**
Maximum	**660**
Sum	1040
Count	4

Total Coliform for Rainwater from Roof

Mean	**405**
Standard Error	159.45
Median	310
Mode	#N/A
Standard Deviation	**318.90**
Sample Variance	101700
Kurtosis	2.14
Skewness	1.46
Range	720
Minimum	**140**
Maximum	**860**
Sum	1620
Count	4

Table A.10 Calculations for TSS completed in Panama

$$\frac{mg}{L}TSS = \frac{(B-A)\times 1000}{V}$$

Sample Volume (V) (mL): 100	A (g): (Weight of Filter Paper)	B (g): (Weight of Filter Paper + Dried Residue)		
Source	Trial #/Storage Tank #	A Values (g)	B Values (g)	mg/L TSS
Municipal Water (from Sink at STRI lab)	**Trial 1**	0.093	0.097	**40**
		0.094	0.095	**10**
		0.095	0.098	**30**
Rain from Roof	**Trial 1**	0.1	0.104	**40**
		0.091	0.093	**20**
		0.092	0.096	**40**
	Trial 2	0.089	0.092	**30**
		0.088	0.093	**50**
		0.098	0.101	**30**
Rain Stored in Rainwater Storage Tanks	**Storage Tank 1**	0.096	0.096	**0**
		0.098	0.099	**10**
		0.097	0.097	**0**
	Storage Tank 2	0.191	0.193	**20**
		0.169	0.171	**20**
		0.19	0.195	**50**
	Storage Tank 3	0.186	0.188	**20**
		0.181	0.182	**10**
		0.17	0.173	**30**
Rain from Overflow Pipe	**Trial 1**	0.191	0.198	**70**
		0.19	0.196	**60**
		0.17	0.181	**110**

Figure A.10(a–d) Costs of building materials (in Kenyan Shillings)

The following materials were quoted at four stores in Nanyuki, Kenya in January 2006.

Store 1

Chicken Wire

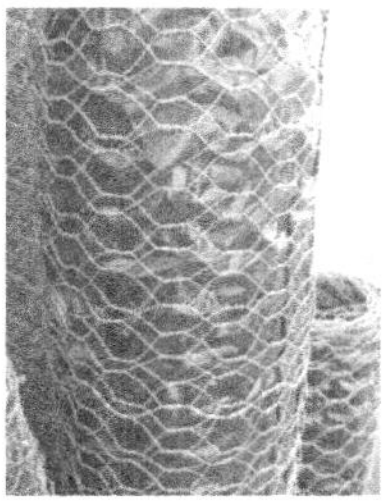

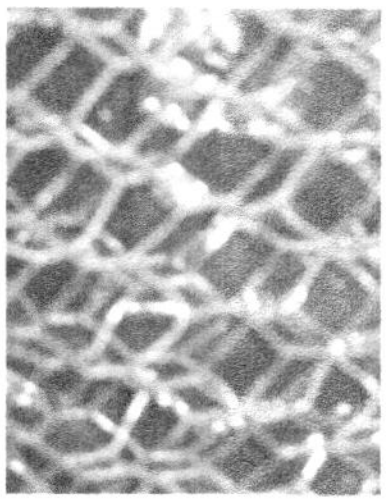
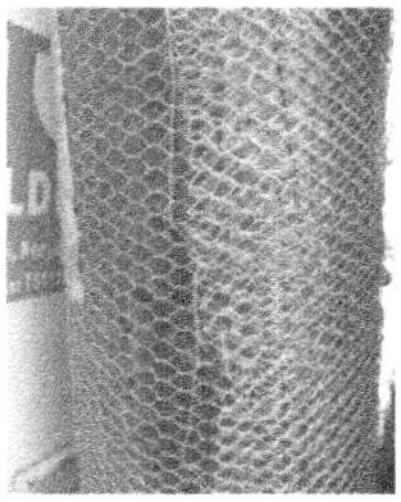

Object	Price (in Kshs)
Chicken Wire (30 m × 1.8 m; 98.4 ft × 6 ft)	2,450.00
Chicken Wire (30 m × 1.8 m; 98.4 ft × 6 ft) [higher strength than (1)]	6,500.00
Chicken Wire (triple twist; 18 m × 1.5 m; 59 ft × 5 ft)	3,250.00

Screens

Screens (per meter)	90.00

Lime

Lime (25 kg)	350.00

Plastic

Clear (per meter)	9.00
Black (per meter) [higher strength than (1)]	80.00
Yellow (per meter) [higher strength than (2)]	90.00

Store 2

Chicken Wire

Object	**Price (in Kshs)**
Chicken Wire (182.9 cm × 3.8 cm; 72 in × 1.5 in)	2,200.00
Chicken Wire (182.9 cm × 2.54 cm; 72 in × 1.0 in)	2,700.00

Wire

Wire (18 m × 1.2 m; 59 ft × 4 ft)	2.500.00
Wire (18 m × 1.5 m; 59 ft × 5 ft)	3,000.00
Wire (18 m × 1.8 m; 59 ft × 6 ft)	3,500.00

Screens

Screens (per meter)	100.00

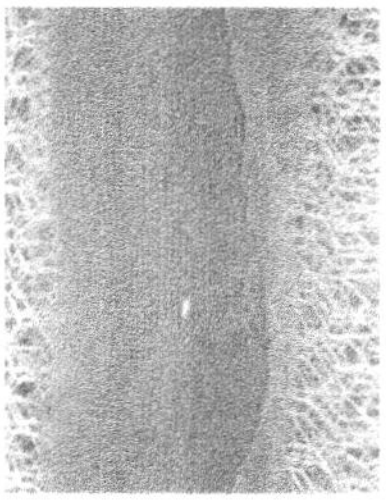

Lime

Lime (25 kg)	250.00
Lime (10 kg)	150.00
Cement (50 kg)	600.00

Galvanized Iron

Galvanized Iron (flat sheets; 6 ft×3 ft; 1.8 m×0.9 m; 32 gage)	400.00
Galvanized Iron (flat sheets; 6 ft×3 ft; 1.8 m×0.9 m; 30 gage)	450.00

Corrugated Iron (3 m×1 m; 9.8 ft×3.3 ft)	450.00

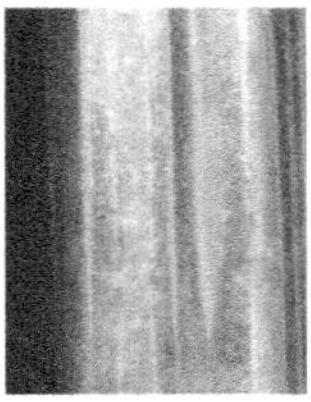

RWH Tanks

HDPE Tank (460 liters; 121.5 gallons)	4,500.00–5,000.00

HDPE Tank (1,000 liters; 264 gallons)	7,000.00
HDPE Tank (2,300 liters; 607.6 gallons)	15,000.00
HDPE Tank (4,600 liters; 1,215 gallons)	30,000.00–30,800.00

Store 3

Object	Price (in Kshs)
Wood (per ft; 8 in wide; 1 in thick)	26.00
Wood (per ft; 12 in wide; 2 in thick)	40.00

Store 4

PVC Pipes

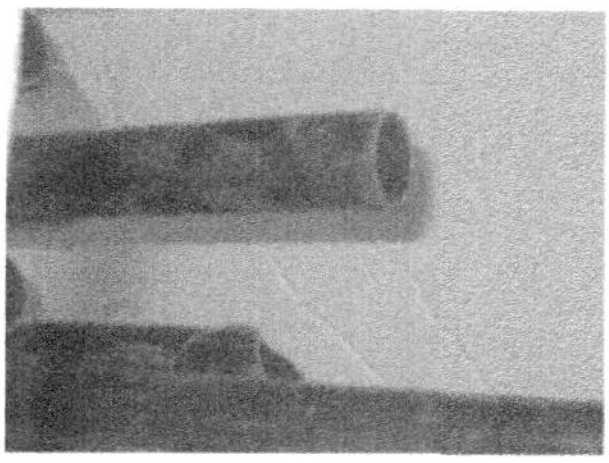

Object	Price (in Kshs)
PVC Pipe (Class A (lightweight); diameter: 1.5 in)	200.00
PVC Pipe (Class A (lightweight); diameter: 2.0 in)	350.00
Galvanized Iron Down-Pipe (length: 6 ft)	120.00

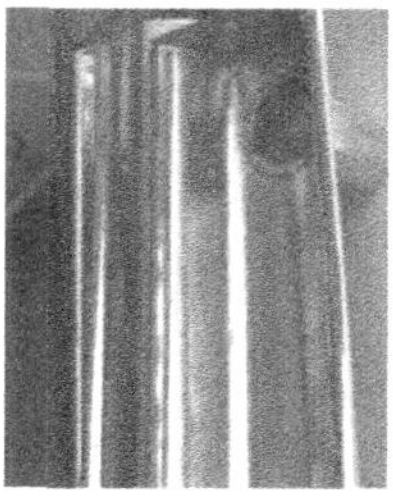

G Bracket	15.00

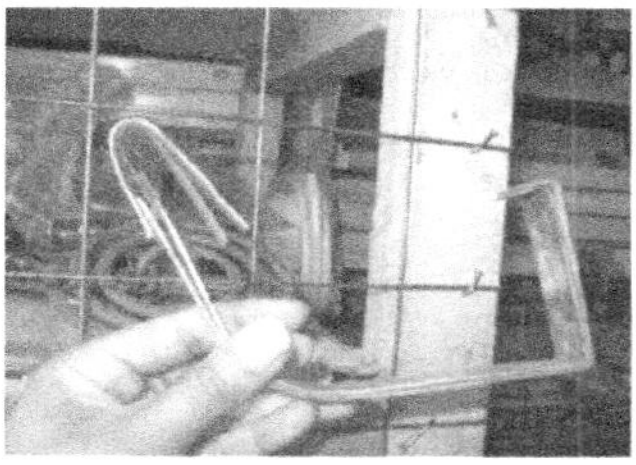

Fascia Boards (6 in wide; 10 ft long; 1 in thick)	50.00

Table A.11 Monthly rainfall data at the Mpala Conservancy and MRC from 1999–2004 (STRI, 2006)

		Conservancy	Centre
YEAR	MONTH	SUM_RAIN (mm)	SUM_RAIN_M (mm)
1999	1	8.4	8.6
1999	2	0.0	0.0
1999	3	104.1	73.9
1999	4	23.1	61.3
1999	5	33.3	36.3
1999	6	1.8	2.1
1999	7	91.9	106.6
1999	8	52.8	61.4
1999	9	14.5	16.8
1999	10	19.6	22.8
1999	11	43.7	50.7
1999	12	15.2	17.7
2000	1	0.3	0.8
2000	2	0.0	0.0
2000	3	0.0	0.0
2000	4	1.8	3.8
2000	5	38.9	32.4
2000	6	24.9	22.7
2000	7	57.2	61.0
2000	8	105.4	148.0
2000	9	6.9	7.5
2000	10	16.8	17.8
2000	11	20.3	33.3
2000	12	7.9	16.6
2001	1	42.4	46.9
2001	2	0.0	0.0
2001	3	50.0	55.7
2001	4	158.0	142.0
2001	5	41.4	94.8
2001	6	62.0	70.8
2001	7	45.0	48.3
2001	8	47.7	54.7
2001	9	6.9	8.5
2001	10	16.8	51.9
2001	11	192.0	232.1
2001	12	8.4	11.0
2002	1	7.9	8.0
2002	2	4.1	0.0

(*continued*)

Table A.11 *Continued*

		Conservancy	Centre
YEAR	MONTH	SUM_RAIN (mm)	SUM_RAIN_M (mm)
2002	3	38.6	43.3
2002	4	96.8	100.2
2002	5	62.7	80.8
2002	6	18.8	16.0
2002	7	34.8	38.0
2002	8	11.4	3.7
2002	9	9.1	9.2
2002	10	48.5	56.0
2002	11	71.1	86.0
2002	12	79.8	91.1
2003	1	0.3	0.0
2003	2	11.2	12.0
2003	3	10.2	6.0
2003	4	224.0	227.0
2003	5	82.3	100.7
2003	6	33.0	36.2
2003	7	34.8	38.8
2003	8	105.4	123.7
2003	9	27.2	30.0
2003	10	11.9	12.0
2003	11	106.9	109.4
2003	12	39.1	48.8
2004	1	78.2	102.5
2004	2	20.1	23.4
2004	3	34.0	41.6
2004	4	166.6	205.3
2004	5	64.5	75.5
2004	6	1.8	2.4
2004	7	78.2	71.2
2004	8	70.1	98.5
2004	9	64.3	0.0
2004	10	32.0	0.0
2004	11	99.6	0.0
2004	12	6.6	0.0

Table A.12 Calculation of storage volume for six buildings in Laikipia, Kenya

Chief of Lekiji's House (5 people × 50 liters of water demanded per day)

Day	Daily Water Demand (m^3)	Rainfall (m)	Roof Size (m^2)	Daily Water Supply (Roof Size∗Rainfall∗0.9)	Storage Volume (m^3)
1	0.25	0.0104	42.5	0.4	0.15
2	0.25	0.0067	42.5	0.26	0.16
3	0.25	0.0103	42.5	0.39	0.3
4	0.25	0.0123	42.5	0.47	0.52
5	0.25	0.0077	42.5	0.3	0.57
6	0.25	0.0058	42.5	0.22	0.54
7	0.25	0.0189	42.5	0.72	1.01
8	0.25	0.0006	42.5	0.02	0.79
9	0.25	0.0008	42.5	0.03	0.57
10	0.25	0.015	42.5	0.57	0.89
11	0.25	0.0098	42.5	0.37	1.01
12	0.25	0.0251	42.5	0.96	1.73
13	0.25	0.0518	42.5	1.98	3.46
14	0.25	0.0038	42.5	0.15	3.35
15	0.25	0.0077	42.5	0.3	3.4
				Maximum Storage Volume (m^3)	**3.46**

Home in Il Motiok (5 people × 50 liters of water demanded per day)

Day	Daily Water Demand (m^3)	Rainfall (m)	Roof Size (m^2)	Daily Water Supply (Roof Size∗Rainfall∗0.9)	Storage Volume (m^3)
1	0.25	0.0104	31.7	0.30	0.05
2	0.25	0.0067	31.7	0.19	−0.01
3	0.25	0.0103	31.7	0.29	0.04
4	0.25	0.0123	31.7	0.35	0.14
5	0.25	0.0077	31.7	0.22	0.12
6	0.25	0.0058	31.7	0.17	0.03
7	0.25	0.0189	31.7	0.54	0.32
8	0.25	0.0006	31.7	0.02	0.09
9	0.25	0.0008	31.7	0.02	−0.14
10	0.25	0.0150	31.7	0.43	0.18
11	0.25	0.0098	31.7	0.28	0.21

(*continued*)

Table A.12 *Continued*

Home in Il Motiok (5 people × 50 liters of water demanded per day)

Day	Daily Water Demand (m^3)	Rainfall (m)	Roof Size (m^2)	Daily Water Supply (Roof Size∗Rainfall∗0.9)	Storage Volume (m^3)
12	0.25	0.0251	31.7	0.72	0.67
13	0.25	0.0518	31.7	1.48	1.90
14	0.25	0.0038	31.7	0.11	1.76
15	0.25	0.0077	31.7	0.22	1.73
				Maximum Storage Volume (m^3)	**1.90**

Note: A conditional statement was added to the Excel Sheet to ensure that negative storage volumes were changed to 0 values so that negative storage numbers were not added to subsequent days.

Lekiji Primary School a (15 people × 6 L of water demanded per day)

Day	Daily Water Demand (m^3)	Rainfall (m)	Roof Size (m^2)	Daily Water Supply (Roof Size∗Rainfall∗0.9)	Storage Volume (m^3)
1	0.09	0.0104	43.6	0.41	0.32
2	0.09	0.0067	43.6	0.26	0.49
3	0.09	0.0103	43.6	0.40	0.81
4	0.09	0.0123	43.6	0.48	1.20
5	0.09	0.0077	43.6	0.30	1.41
6	0.09	0.0058	43.6	0.23	1.55
7	0.09	0.0189	43.6	0.74	2.21
8	0.09	0.0006	43.6	0.02	2.14
9	0.09	0.0008	43.6	0.03	2.08
10	0.09	0.0150	43.6	0.59	2.58
11	0.09	0.0098	43.6	0.38	2.87
12	0.09	0.0251	43.6	0.99	3.77
13	0.09	0.0518	43.6	2.03	5.71
14	0.09	0.0038	43.6	0.15	5.77
15	0.09	0.0077	43.6	0.30	5.99
				Maximum Storage Volume (m^3)	**5.99**

(*continued*)

Table A.12 *Continued*

Lekiji Primary School b (25 people × 6 L of water demanded per day)

Day	Daily Water Demand (m^3)	Rainfall (m)	Roof Size (m^2)	Daily Water Supply (Roof Size∗Rainfall∗0.9)	Storage Volume (m^3)
1	0.15	0.0104	71.4	0.67	0.52
2	0.15	0.0067	71.4	0.43	0.80
3	0.15	0.0103	71.4	0.66	1.31
4	0.15	0.0123	71.4	0.79	1.95
5	0.15	0.0077	71.4	0.50	2.30
6	0.15	0.0058	71.4	0.38	2.53
7	0.15	0.0189	71.4	1.22	3.59
8	0.15	0.0006	71.4	0.04	3.48
9	0.15	0.0008	71.4	0.05	3.38
10	0.15	0.0150	71.4	0.96	4.20
11	0.15	0.0098	71.4	0.63	4.68
12	0.15	0.0251	71.4	1.62	6.14
13	0.15	0.0518	71.4	3.33	9.32
14	0.15	0.0038	71.4	0.24	9.42
15	0.15	0.0077	71.4	0.50	9.76
				Maximum Storage Volume (m^3)	**9.76**

Lekiji Primary School c (60 people × 6 L of water demanded per day)

Day	Daily Water Demand (m^3)	Rainfall (m)	Roof Size (m^2)	Daily Water Supply (Roof Size∗Rainfall∗0.9)	Storage Volume (m^3)
1	0.36	0.0104	91.6	0.86	0.50
2	0.36	0.0067	91.6	0.55	0.69
3	0.36	0.0103	91.6	0.85	1.18
4	0.36	0.0123	91.6	1.02	1.84
5	0.36	0.0077	91.6	0.64	2.12
6	0.36	0.0058	91.6	0.48	2.24
7	0.36	0.0189	91.6	1.56	3.44
8	0.36	0.0006	91.6	0.05	3.13
9	0.36	0.0008	91.6	0.06	2.83
10	0.36	0.0150	91.6	1.24	3.71
11	0.36	0.0098	91.6	0.81	4.15

(*continued*)

Table A.12 *Continued*

Lekiji Primary School c (60 people × 6 L of water demanded per day)

Day	Daily Water Demand (m^3)	Rainfall (m)	Roof Size (m^2)	Daily Water Supply (Roof Size∗Rainfall∗0.9)	Storage Volume (m^3)
12	0.36	0.0251	91.6	2.07	5.87
13	0.36	0.0518	91.6	4.27	9.78
14	0.36	0.0038	91.6	0.31	9.73
15	0.36	0.0077	91.6	0.64	10.01
				Maximum Storage Volume (m^3)	**10.01**

Lekiji Teacher Homes (8 people × 50 L of water demanded per day)

Day	Daily Water Demand (m^3)	Rainfall (m)	Roof Size (m^2)	Daily Water Supply (Roof Size∗Rainfall∗0.9)	Storage Volume (m^3)
1	0.4	0.0104	60.2	0.56	0.16
2	0.4	0.0067	60.2	0.36	0.13
3	0.4	0.0103	60.2	0.56	0.29
4	0.4	0.0123	60.2	0.67	0.55
5	0.4	0.0077	60.2	0.42	0.57
6	0.4	0.0058	60.2	0.32	0.49
7	0.4	0.0189	60.2	1.03	1.12
8	0.4	0.0006	60.2	0.03	0.75
9	0.4	0.0008	60.2	0.04	0.39
10	0.4	0.0150	60.2	0.81	0.80
11	0.4	0.0098	60.2	0.53	0.93
12	0.4	0.0251	60.2	1.36	1.90
13	0.4	0.0518	60.2	2.81	4.30
14	0.4	0.0038	60.2	0.21	4.11
15	0.4	0.0077	60.2	0.42	4.13
				Maximum Storage Volume (m^3)	**4.30**

Table A.13 MRC daily rainfall data for 2001 and 2003 starting at average values of the 15-day consecutive rainfall events in November 2001 and April 2003 (STRI, 2006)

Day	November 2001 Rainfall (mm)	April 2003 Rainfall (mm)	Day	November 2001 Rainfall (mm)	April 2003 Rainfall (mm)
1	**10.416**	**10.416**	26	0.254	5.588
2	**6.729**	**6.729**	27	0.000	0.000
3	**10.286**	**10.286**	28	7.110	0.000
4	**12.32**	**12.32**	29	0.000	3.048
5	**7.747**	**7.747**	30	0.000	1.524
6	**5.842**	**5.842**	31	0.000	0.254
7	**18.924**	**18.924**	32	0.000	0.000
8	**0.635**	**0.635**	33	0.000	12.950
9	**0.762**	**0.762**	34	0.000	0.000
10	**14.987**	**14.987**	35	0.000	0.000
11	**9.781**	**9.781**	36	0.000	2.032
12	**25.148**	**25.148**	37	0.000	4.826
13	**51.815**	**51.815**	38	0.000	0.000
14	**3.809**	**3.809**	39	0.000	0.254
15	**7.745**	**7.745**	40	0.000	0.000
16	0.000	0.000	41	0.000	0.508
17	7.110	8.130	42	0.762	0.000
18	4.826	29.210	43	0.000	0.000
19	5.080	3.302	44	7.110	0.000
20	0.000	0.000	45	0.508	0.000
21	0.000	0.000	46	0.000	0.000
22	10.160	0.762	47	0.000	0.000
23	0.000	0.254	48	0.000	0.000
24	0.254	1.270	49	0.000	0.000
25	9.910	0.000	50	0.000	0.000

Note: Chart Only Includes 35 Days of Subsequent Rainfall in 2001 and 2003.

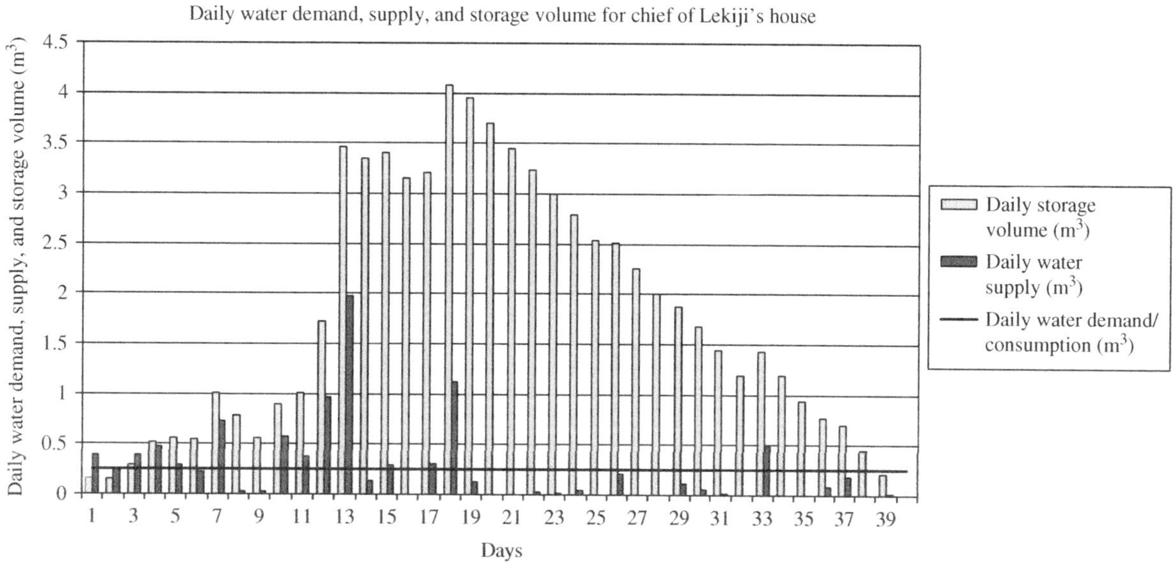

Figure A.11.a Daily water demand, supply, and storage volume for a house in Lekiji
(Starting from significant rainfall event in *April 2003* and continuing until daily storage volume is zero or less)

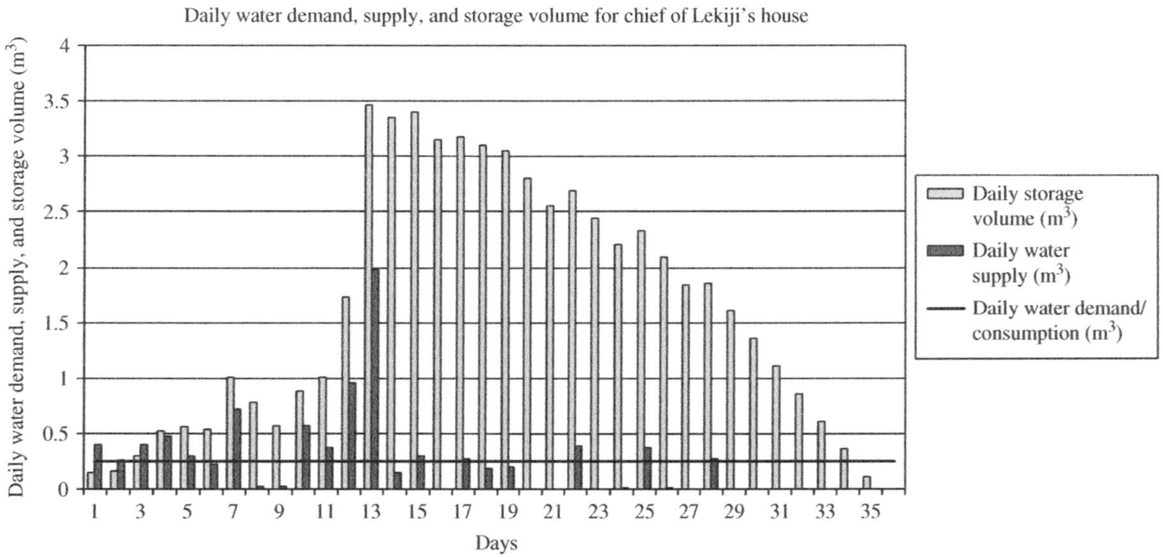

Figure A.11.b Daily water demand, supply, and storage volume for a house in Lekiji (Starting from significant rainfall event in *November 2001* and continuing until daily storage volume is zero or less)

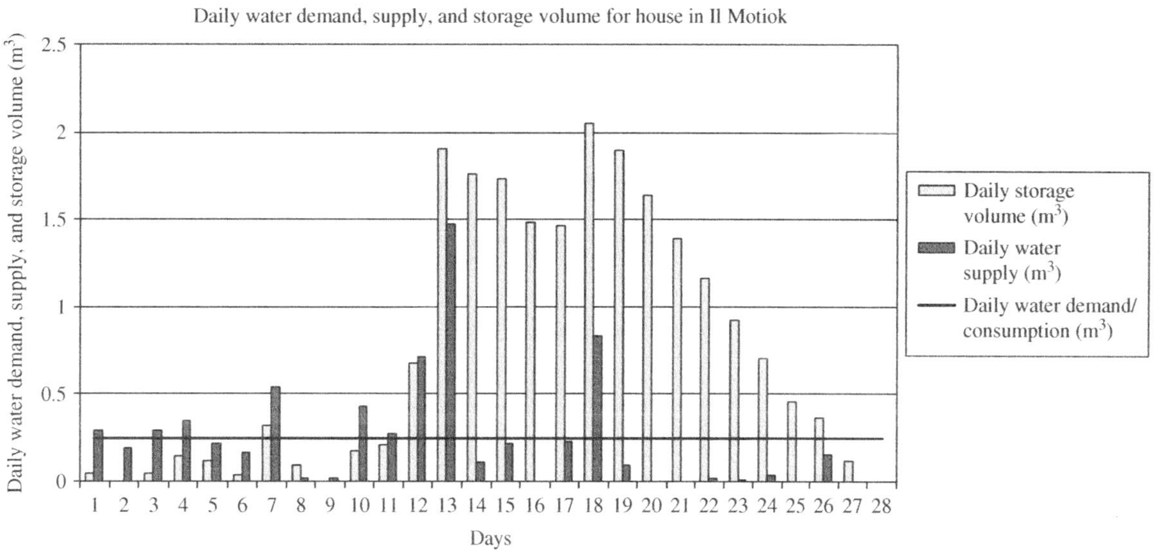

Figure A.11.c Daily water demand, supply, and storage volume for a house in Il Motiok
(Starting from significant rainfall event in *April 2003* and continuing until daily storage volume is zero or less)

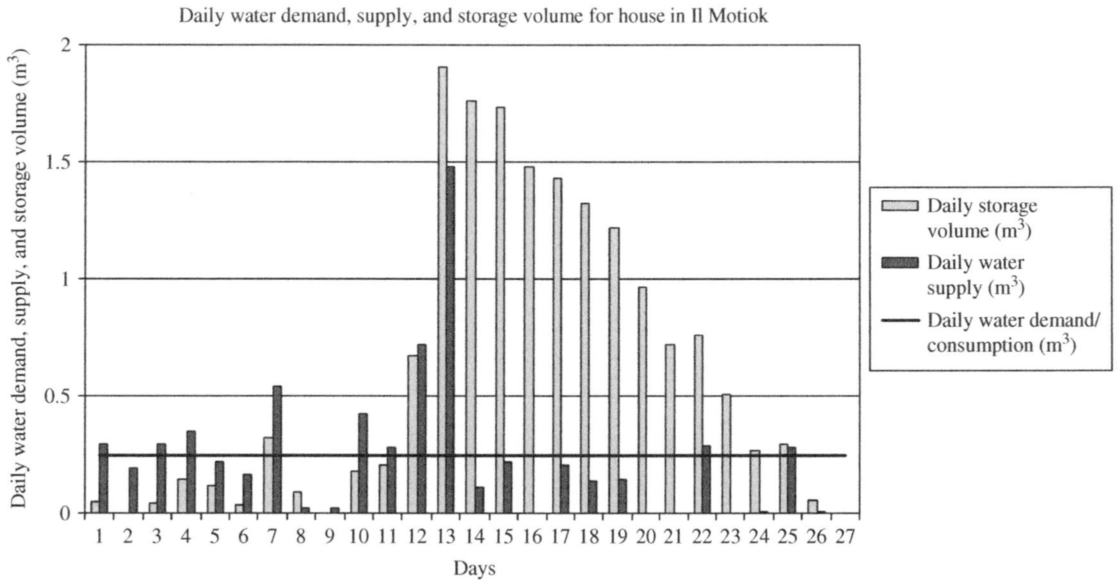

Figure A.11.d Daily water demand, supply, and storage volume for a house in Il Motiok
(Starting from significant rainfall event in *November 2001* and continuing until daily storage volume is zero or less)

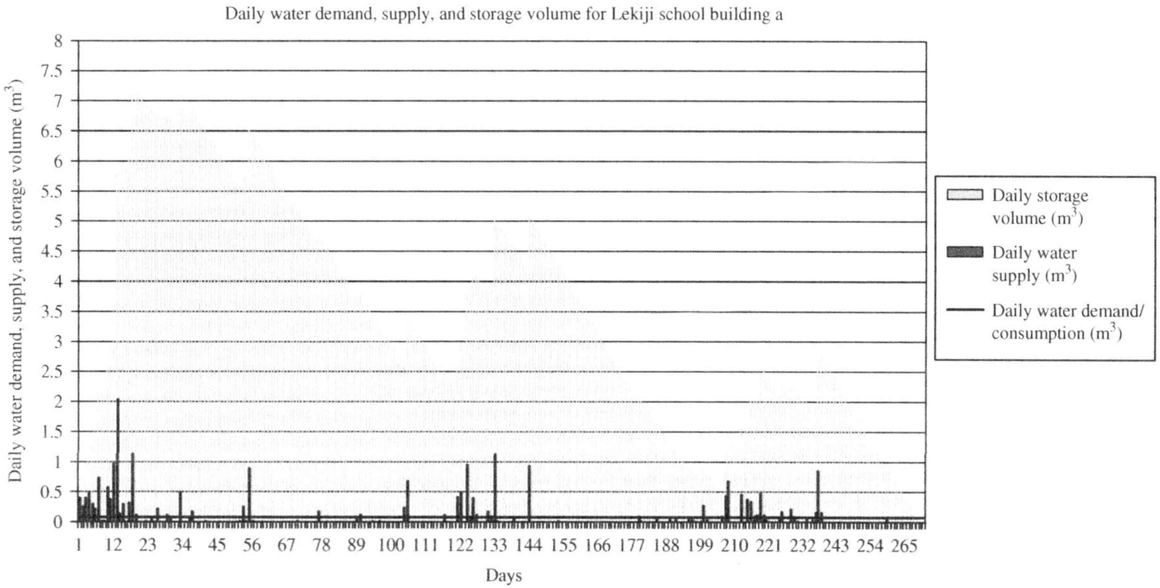

Figure A.11.e Daily water demand, supply and storage volume for Lekiji Primary School a (Starting from significant rainfall event in *April 2003* and continuing until daily storage volume is zero or less)

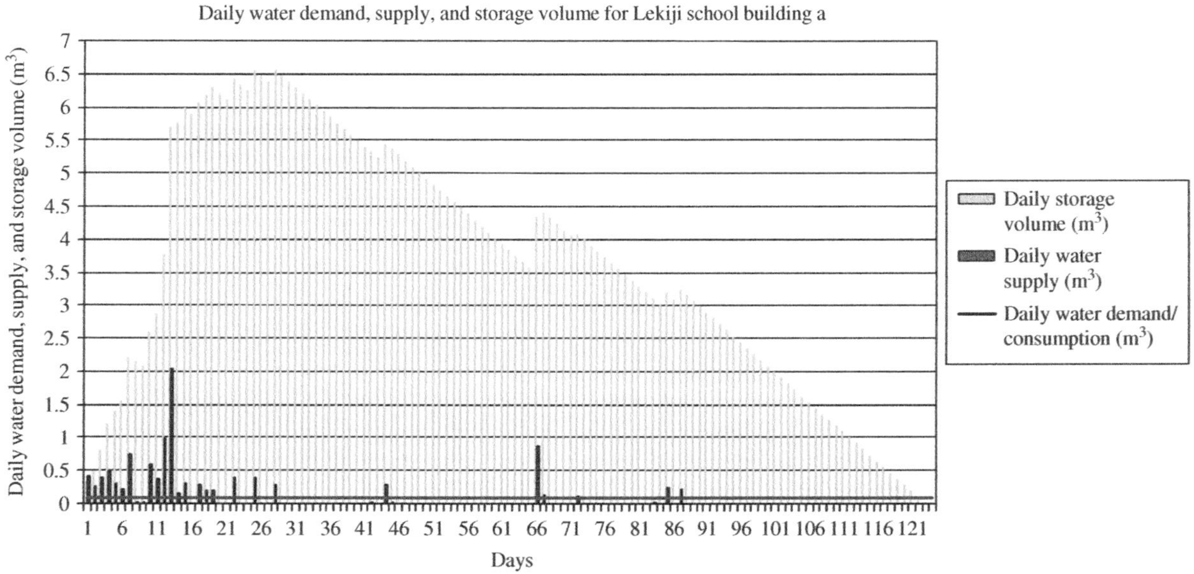

Figure A.11.f Daily water demand, supply, and storage volume for Lekiji Primary School a (Starting from significant rainfall event in *November 2001* and continuing until daily storage volume is zero or less)

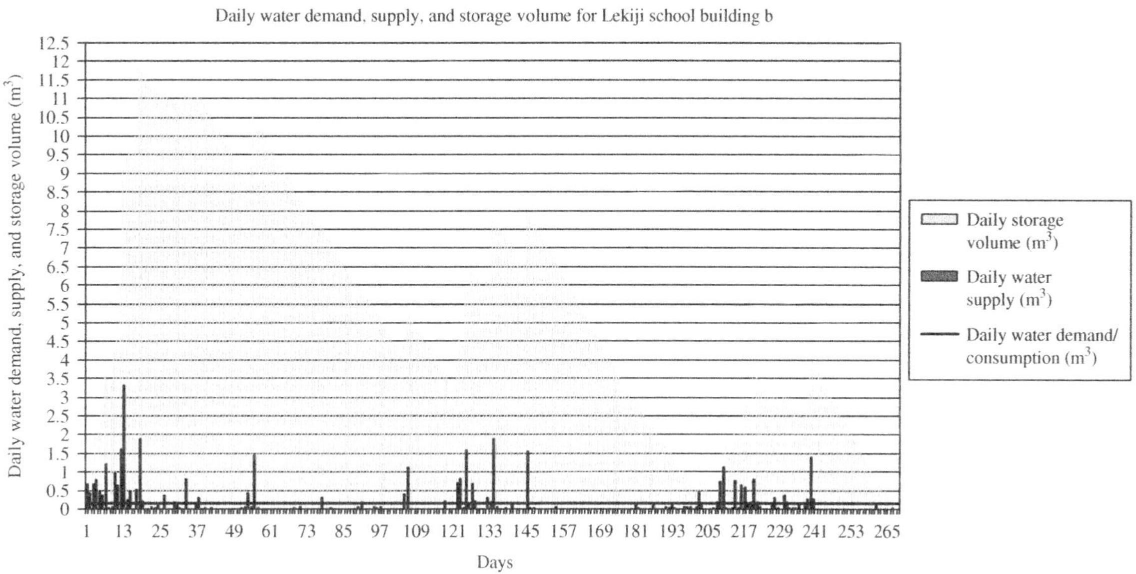

Figure A.11.g Daily water demand, supply, and storage volume for Lekiji Primary School b (Starting from significant rainfall event in *April 2003* and continuing until daily storage volume is zero or less)

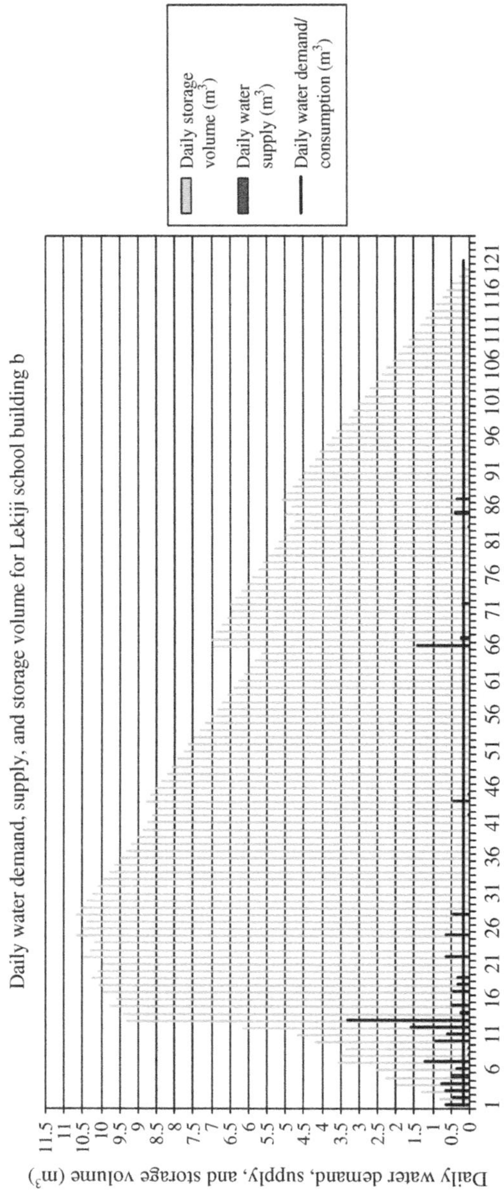

Figure A.11.h Daily water demand, supply, and storage volume for Lekiji Primary School b (Starting from significant rainfall event in *November 2001* and continuing until daily storage volume is zero or less)

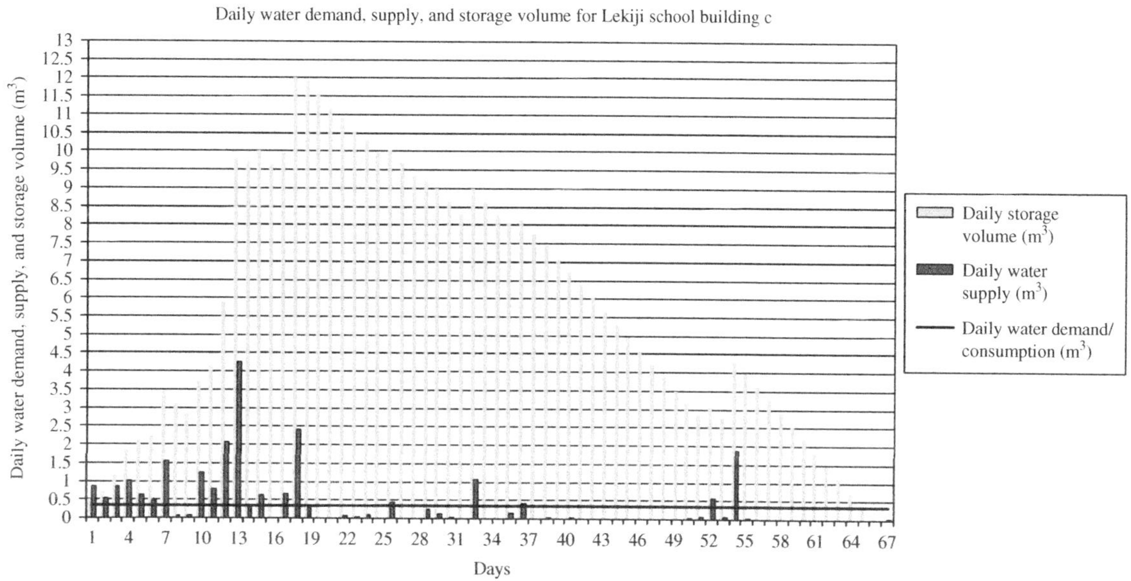

Figure A.11.i Daily Water demand, supply, and storage volume for Lekiji Primary School c (Starting from significant rainfall event in *April 2003* and continuing until daily storage volume is zero or less)

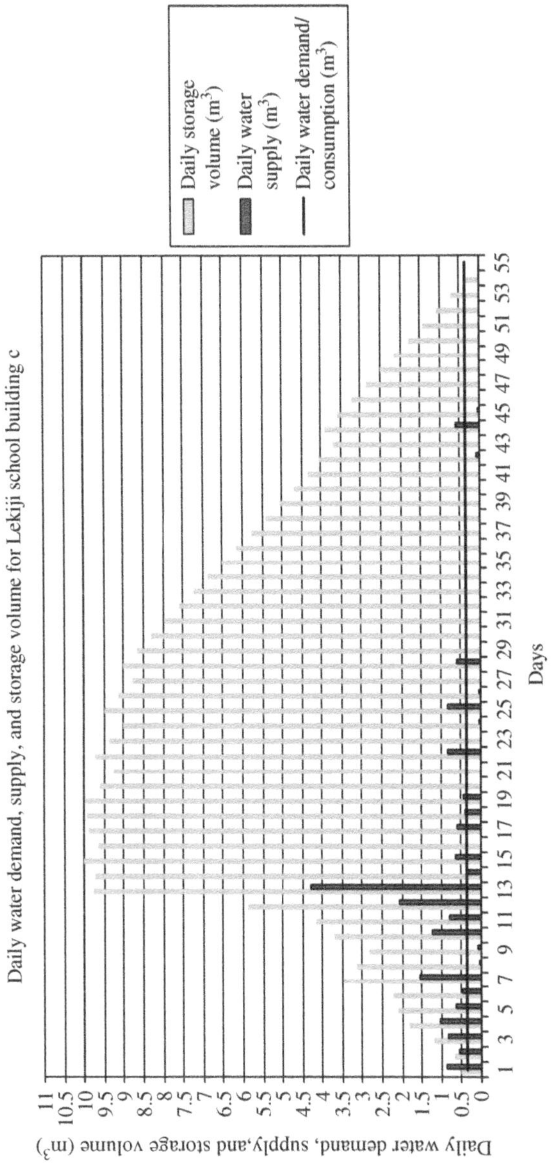

Figure A.11.j Daily water demand, supply, and storage volume for Lekiji Primary School c (Starting from significant rainfall event in *November 2001* and continuing until daily storage volume is zero or less)

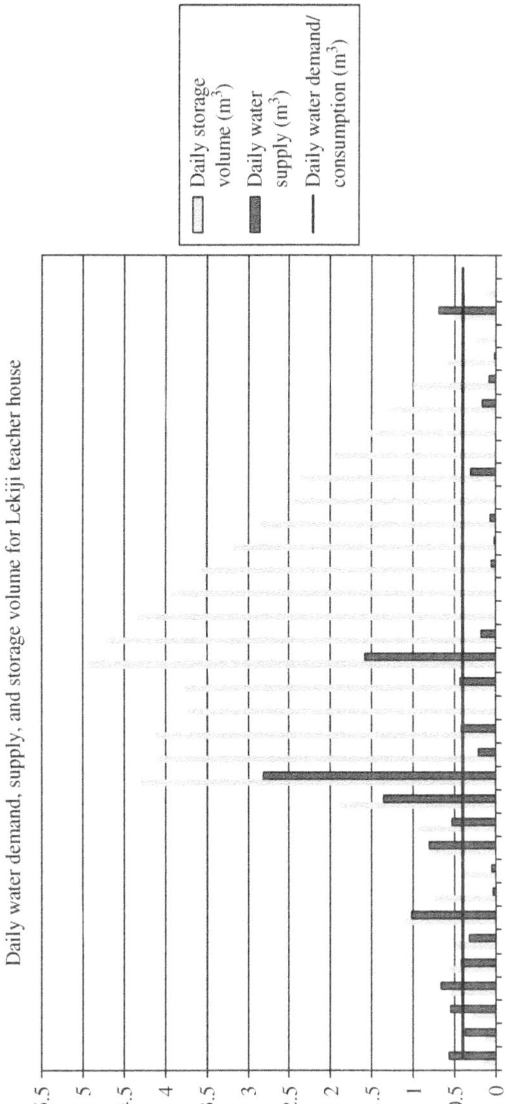

Figure A.11.k Daily water demand, supply, and storage volume for Lekiji teacher house (Starting from significant rainfall event in *April 2003* and continuing until daily storage volume is zero or less)

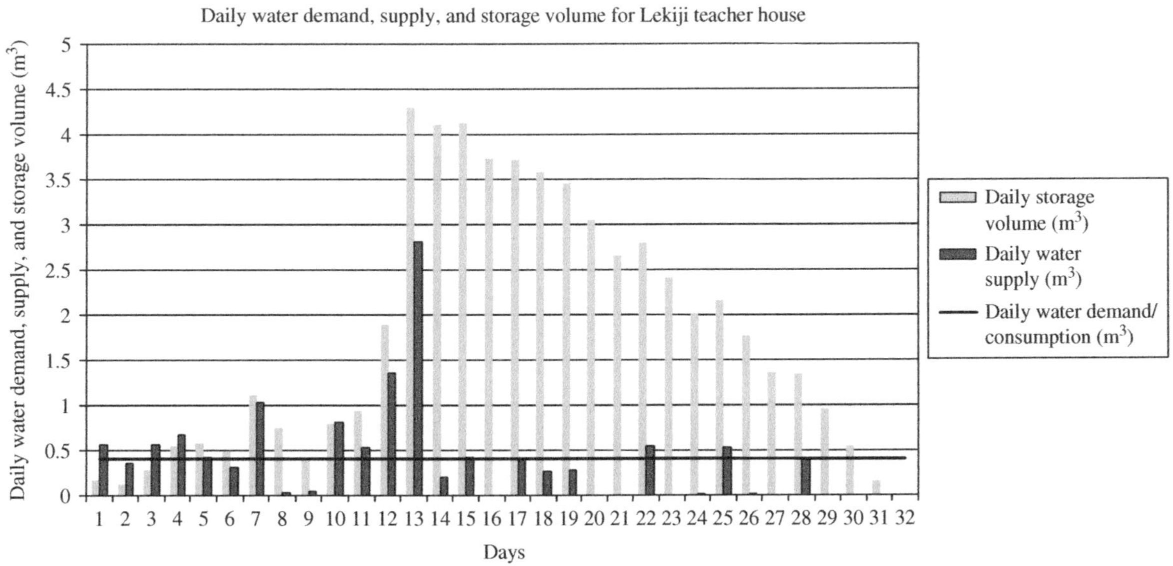

Figure A.11.I Daily water demand, supply, and storage volume for Lekiji teacher house (Starting from significant rainfall event in *November 2001* and continuing until daily storage volume is zero or less)

Subject index

ISBN: 9781843393238. Published by IWA Publishing, London, UK.

IWA Publishing's authorised EU representative for General Product
Safety Regulations is Diane D'Arras, 15 rue Duret, 75116 Paris,
France, e-mail: safety@iwap.co.uk.

Printed and bound by CPI Group (UK) Ltd, Croydon, CR0 4YY
16/06/2026
02140394-0003